Analysis Using Glass Electrodes

Analysis Using Glass Electrodes

Peter W. Linder and Ralph G. Torrington
Department of Physical Chemistry, University of Cape Town

AND

David R. Williams
Department of Applied Chemistry, University of Wales Institute of Science and Technology, Cardiff

Open University Press
A division of
Open University Educational Enterprises Ltd.
12, Cofferidge Close
Stony Stratford
Milton Keynes MK11 1BY, England

Copyright © Open University Press 1984

All rights reserved. No part of this work may be reproduced in any form, by mimeograph or by any other means, without permission in writing from the publisher.

British Library Cataloguing in Publication Data

Linder, P. W.
   Analysis using glass electrodes.
   1. Electrochemical analysis   2. Electrodes, Glass
   I. Title   II. Torrington, R. G.   III. Williams, D. R.
   543'.0871   QD115

ISBN 0-335-10420-7

Text design by Clark Williams
Cover design by Phil Atkins

Printed in Northern Ireland at The Universities Press (Belfast) Ltd.

Contents

Foreword by Professor R. G. Bates, University of Florida   ix

Preface   x

Acknowledgements   xii

Chapter 1. Introduction   1

    History of development of pH concept   2
    Development of the glass electrode   3
    Definitions of pH   4
    Activity versus concentration   5
    The pH scale   6
    pH measurements   7
    pH standards   8
    Millivolt approaches for hydrogen-ion concentrations   10
    Electrode millivolt calibrations   10
    Optimization techniques   11
    Formation constants   11
    Speciation, bioavailability, and the scheme of this book   12

Chapter 2. Design and maintenance of electrodes   13

    Construction   13
    Glass composition   15
    Robustness   16

Filling solutions	16
Reference electrodes	16
Combination electrodes	18
Commissioning a new electrode	18
Electrode storage	19
Cleaning and transferring electrodes	19
Meters	20
Glass electrodes for special purposes	20

Chapter 3. Determination of pH values — 22

Introduction	22
Activities and activity coefficients	23
Operational definition of pH	27
Definition of the pH scale	28
Measurements of pH in practice	32
Principles of buffer action	33
Recipes for standard buffers	38
Procedures for routine pH measurements	38
pH measurements in solvents other than water	40
The interpretation of pH	41
Systems of high ionic strength	41
Concluding remarks	42

Chapter 4. Determination of hydrogen-ion concentrations — 44

Data	44
Strong-acid–strong-base procedures	46
Internal calibration procedures	47
Assessment of analytical errors	48
Programs available	49
MAGEC	49
General	49
Principles of MAGEC	50

MAGEC in practice	51
Worked examples using MAGEC and MAGEC–MINIQUAD cycling	52
Discussion of MAGEC	57
ACBA	57
General	57
Principles involved in ACBA	57
Worked examples using ACBA	58
Discussion of ACBA	61
LETAGROP	61
General	61
Principles involved in LETAGROP	61
Examples using LETAGROP	62
MINIPOT	62
General	62
Principles involved in the MINIPOT algorithm	62
Worked examples using MINIPOT	63
Discussion of MINIPOT	63
LIGEZ	63
General	63
Principles involved in LIGEZ	64
Worked examples using LIGEZ	64
Discussion of LIGEZ	65
The experimental determination of formation constants	68
Concluding remarks and our experiences	68

Chapter 5. Determination of complex concentrations and speciation 71

Input data required for a speciation calculation	71
Applications of speciation studies	74
Complex-formation research	75
Medicine	76
Industrial uses	79
The oceans and the atmosphere	82
Environmental problems	84
Bioavailability of metals as toxins or nutrients	84
Lessons learned, and their application to other ion-selective electrodes	85
Concluding remarks	85

Appendices. Computer programs referred to in the text   **87**

A	The MAGEC program	87
	Data input	87
	Program listing	88
B	The ACBA program	111
	Data input	111
	Program listing	112
C	The MINIPOT program	124
	Program listing	124
	Numerical examples	129
D	The LIGEZ program	132
	Theory	132
	The LIGEZ program coding	135
	LIGEZ coding for HP41C calculator	135
	Running the LIGEZ program	138
	Program execution	139

References   141

Index   145

Foreword

In the earliest years of this century, analytical potentiometry, a branch of applied electrochemistry, was limited largely to potentiometric titrations, utilizing as indicator-electrodes the hydrogen electrode and a few redox systems. In defining the pH unit in 1909, Sørensen provided a solution parameter that has been of inestimable impact on commerce and industry as well as in chemical analysis and the clinical laboratory. For some 30 years, however, the inconvenience and limitations of the hydrogen electrode frustrated all attempts to extend the application of pH measurements to a wide variety of areas of practical interest.

Although the hydrogen-ion response of glass membranes had been observed much earlier, it was not until the 1930s that glass electrodes of practical utility became a reality. The high electrical resistance of these sensors was still a barrier to their general application, however, until the development of suitable electronic instrumentation in the next decade. These advances together promoted a notable renaissance of potentiometry as an analytical tool when pH determinations became commonplace. The meaning of these pH numbers obtained so easily was, and continues to be, obscure and of less concern than their reproducibility.

An analysis of the response of the common pH cell demonstrates that the potential of the glass electrode is a function of hydrogen-ion activity (a_H) under ideal measurement conditions. Reference standards for pH based on a conventional (arbitrary) scale of a_H provide the required reproducibility and, in restricted circumstances, allow a useful interpretation of pH in terms of the thermodynamic properties of ionic solutions. Nevertheless, the concept of hydrogen-ion concentration, as opposed to conventional activity, is clear and well understood. An experimental measure of hydrogen-ion concentrations is needed and, happily, is often accessible. It is to be hoped that an understanding of the behaviour of glass electrodes in constant ionic media may serve to promote a wider application of pH measurements in solution chemistry. This book, appearing appropriately near the 75th anniversary of the pH unit, brings us a step closer to this worthwhile goal.

Gainesville, Florida, USA ROGER G. BATES
November 1983

Preface

1984 is the 75th anniversary of the discovery of pH. Since 1909 many techniques for measuring pH and for researching the basic underlying theories have been developed. The utilization of this increased knowledge has given rise to new manufacturing technologies for glass electrodes and for millivoltmeters and so, whenever the chemical industries or analytical sciences use aqueous solutions, glass-electrode measurements play a central role in process control and the maintenance of standards.

Over the years, several excellent books and reviews have appeared concerning pH theory and practical aspects but after such a long era of pH 'the time is surely ripe to feel our way towards measuring hydrogen ion concentrations from the emf of pH cells' (Professor R. G. Bates). This book aims to point the way forward towards such a new generation of glass-electrode measurements.

P.W.L.
R.G.T.
D.R.W.

November 1983

Acknowledgements

We would like to express our sincere thanks to Mrs Phyllis Bevan for the typing, Mrs Susan L. Abraham for artwork, Drs Guiseppe Arena, Fabri Marsicano, Peter May, and Silvio Sammartano for providing details for the manuscript, to our families for their indulgence and support, and to our universities for providing research facilities.

We gratefully acknowledge contributions of photographs from Metrohm AG (Switzerland) and Radiometer A/S (Denmark)—Figures 3.1 and 2.1 respectively; permission to reproduce copyright material from Pergamon Press Ltd. (England)—ACBA, MAGEC and MINIPOT programs; and Mintek (South Africa) for Figure 5.5.

P.W.L.
R.G.T.
D.R.W.

1

Introduction

The glass electrode is to solution chemistry what the silicon chip is to computers—the central component. This book has as its three objectives descriptions of pH, hydrogen ion concentration, [H$^+$], and complex speciation measurements all using glass electrodes and electrometric monitoring.

As with the 'chip', the glass electrode is dependent upon supporting instrumentation such as voltmeters and reference electrodes. However, enhancements in the accuracy and predictability of the values arising from glass electrode potentiometry require a full mastery of the principles behind electrode calibrations. Thus, we make no excuses for placing calibration theories for both pH and [H$^+$] in the central role of this book. On the one hand we aim to update descriptions of pH measurements and, on the other hand, we shall show how computer programs and new theories of optimization now make possible precise free-ion concentration measurements which opens up new vistas in 'speciation' assessment.

It is equally possible to use glass electrodes to measure hydrogen-ion *concentrations*, expressed as [H$^+$], or *activities*, expressed as pH (Bates 1954, 1973, 1981; May *et al.* 1982). Many scientists are 'bilingual' in that they have a mastery of both concepts. The choice of language used frequently depends upon why the measurements are required: [H$^+$] measurements are needed in various specialized areas of chemistry (pure and applied) and biochemistry, involving, for example, determination of equilibrium constants, rate constants, and the concentration distributions of interacting species in a system (called speciation studies).

Measured values of pH provide a 'characteristic number' giving a relative measure of acidity. Reproducibility of pH between different laboratories or at different times is often of greater importance than its exact physical significance. In spite of inherent difficulties in attaching physical significance to recorded pH values, the work of various scientists (see those mentioned in Bates 1981; Covington 1981) and bodies (British Standards Institution, US National Bureau of Standards, International Union of Pure and Applied

Chemistry) leads to definitions which provide the best possible interpretation of these values in terms of hydrogen-ion activities. Several methods of measuring pH have been devised: the most *exact* utilizes a hydrogen electrode; the most *important* routine method utilizes a glass electrode.

The 'second language' of acidity—that which talks in terms of hydrogen-ion concentrations, [H^+]—has been less well documented in books and considered by international bodies of scientists until quite recently. Compared with pH, which has passed from one meaning to another as our understanding expanded, there is no ambiguity about the meaning of [H^+]. Further, experimental and computational techniques for determining [H^+] from glass-electrode emf measurements are now available. Unfortunately, much of this information lies in research papers on speciation and in books on complex chemistry. The equipment used is the same as that used for pH measurements, but the theory differs.

This book surveys the literature concerning pH, and indicates the most modern meanings of pH values and means of measuring them; next it presents a full description of the best methods available for measuring [H^+], and then it illustrates the extreme usefulness of [H^+] in permitting formation constant, and thence speciation measurements. (It is assumed that the reader is already convinced of the usefulness of pH measurements.)

History of development of pH concept

In 1909, Sørensen, in Denmark, researching acid-dependent enzymatic processes in the brewing industry, noted the difference between the total acid added to a system and the *degree of acidity*. He proposed standard approaches to measuring the useful hydrogen ion left in a system compared with the total acid added and employed both colourimetric and electrometric methods to measure the *power, Potenz* or *puissance* of this hydrogen ion denoted, in those days, as p_H. This Sørensen scale has long since evolved into more sophisticated versions, but all pH measurements pay tribute to the foresight and understanding of this distinguished Danish scientist.

At that time, the far-reaching influence of Sørensen's proposals in industry and medicine were unimaginable and were only realized when the hydrogen ion function of glass membranes was made available in the form of robust glass electrodes and matching voltmeters. As glass electrode—reference-electrode potentials became measurable with speed and accuracy, so too several pH scales 'all masquerading under the name pH' came into common use (Bates 1981). More recently, automation has introduced the possibilities of holding pH values of solutions constant and, for example, keeping acidity conditions optimized for the type of enzyme processes researched by Sørensen.

As in other sciences, the technology often outstrips theory; and so the science of producing numbers from volt- or pH-meters originally surged ahead of our understanding of the underlying theories, and even of the meaning of the numbers. There is no blame implied in this statement since

Introduction

solution thermodynamics was brought to maturity long after Sørensen's pH discoveries.

We now know that a pH 'number' primarily reflects the *activity* rather than the *concentration* of hydrogen ions in a test solution. Furthermore, this activity is based on other ions present as well as H^+ with respect to some internationally agreed reference buffer solution(s) of assigned pH 'number(s)'.* There is still no general agreement concerning which conventional buffer references to adopt. However, the increasing adoption of standardized practical approaches has led to most groups in different countries adopting operational definitions of practical pH values and/or of standard pH scales fixed with respect to standard buffer solutions having assigned pH values. These buffer-solution-assigned pH values match thermodynamic estimates of corresponding H^+ ionic activities as closely as possible.

Thus, any pH 'number' measured is only as good as the standard pH used to set up the experiment. Moreover, the interpretation is limited by our inability to measure activities of single ionic species. For much industrial/biological monitoring this is perfectly adequate but, in research laboratories, there is an increasing need to measure absolute hydrogen-ion *concentrations*. Both equipment and theory are now well able to do this with the precision required.

To maximize the usefulness acquired from analyses using glass electrodes requires a mastery of many features. This volume takes us from glass electrodes, to electrode pairs, to potentiometric techniques, to pH, and then millivolt measurements to $[H^+]$, to chemical equilibrium constants and to speciation. Since each of these factors depends upon the preceding ones, it is axiomatic that one has to enquire most painstakingly into the precise nature of each quantity embodied in the measurements, at all times. Furthermore, objective least-squares decisions ought to be used for data processing wherever possible.

Development of the glass electrode

In 1875 Lord Kelvin suggested that glass was an electrolytic conductor; then Cremer, a generation later in 1906, reported an emf across a thin glass membrane separating two aqueous solutions and noted that this glass-electrode potential was sensitive to changes in acidity. Haber and Klemensiewicz quantified the dependence of this effect with respect to $[H^+]$ in 1909. The transformation of this concept into a practical pH-measuring device was perfected by Dole, Hughes and MacInnes in the 1920s and 1930s (Dole 1941; Mattock and Taylor 1961; Eisenmann 1967).

Ideally, glass membranes ought to show a perfect Nernstian response to

---

* In 1900, Fernbach and Hubert noted that partly neutralized solutions of phosphoric acid tended to oppose changes in hydrogen ion concentration, and compared this resistance to that exerted by '*un tampon*'. This term was used by Sørensen, and translated into German as '*puffer*', and then into English as 'buffer'.

pH. It is well known that no glass gives a 'true' response; not only $E^\ominus$ but also the Nernstian slope tends to be variable and so each electrode has to be calibrated either using buffers to read pH or standard acid and alkali solutions to read emf ($E^\ominus$ is defined on p. 8.)

The glasses used for monitoring different solutions vary. Pyrex or silica glasses having no alkali metal content have very poor pH properties; adding alkali metals introduces a pH response. Surface hygroscopicity is also important—the surface of the glass must be hydrated to produce the pH-dependent emf. Standard MacInnes–Dole glass (a eutectic soda-lime–silica mixture) attracts water molecules to solvate the cations at the surface. Lithium containing glasses require less hydration and so are more useful in partially aqueous media.

We shall not go into great depth concerning glass electrode design since one rarely finds a scientist making his own electrodes nowadays; instead they are purchased from a range of suppliers. Thus, we concentrate more on electrode assessment and calibration in order to advise the reader how best to choose, to test and to use electrodes.

As glass-electrode usage became more widespread, it became apparent that they respond to ions other than H^+. On the one hand this has led to a whole range of electrodes designed to respond selectively to a family of ions and, on the other hand, specificity factors for an ion and the interference effects of other ions are now much better understood and taken into account.

Definitions of pH

The principles of pH measurement commence with a definition of the term pH. The p means power; the H is the symbol for hydrogen. Thus, the term pH represents the hydrogen-ion exponent.

The pH of a solution is a relative measure of its acidity. It is based upon the fundamental concept that

$$pH = -\lg a_{H^+} \text{ or } 10^{-pH} = a_{H^+} \qquad (1.1)$$

but in practice difficulties arise through our inability to measure single-ion activities and so, nowadays, it is defined in an operational manner which is discussed in detail in Chapter 3. This activity is the effective concentration of the hydrogen ion present in solution. Essentially, the difference between effective and actual concentration decreases when moving towards more dilute solutions in which interactions between adjacent ions become progressively less important. It is important to note that for every decade change in activity, the pH changes by one unit. (Throughout this book, whenever reference is made to the hydrogen ion, strictly it ought to be made to the hydronium ion. It is a matter of convention and for brevity that only the hydrogen ion is mentioned even though it is always present as its solvated form.)

Introduction

Activity versus concentration

pH glass electrodes respond to hydrogen-ion activity, a_{H^+}, so the factors which influence activity and its definition are of primary importance. The activity of the hydrogen ion is defined with respect to concentration $[H^+]$ and the activity coefficient, y_{H^+}:

$$a_{H^+} = y_{H^+}[H^+]. \qquad (1.2)$$

There are three commonly used concentration scales—molar, molal and mole fraction. As might be expected, their respective activity coefficients are slightly different for the same solution, but parallel each other in all known trends.

Activity coefficients approximate to unity in dilute solutions, where the ionic strength is low. Thus, the activity is related to concentration through a salt effect, y^x, and a solvent or medium effect, y^m, the glass-electrode measurement of activity being mainly influenced by the ionic strength, the temperature, and the solvent:

$$a_{H^+} = y_{H^+}^x y_{H^+}^m \cdot C_{H^+}. \qquad (1.3)$$

Activity coefficient, ionic strength and pH are interrelated and so the sample temperature, ionic strength and solvent composition should be unambiguously stated when reporting a pH value if another person is to duplicate the results, or if pH values are to be compared.

Factors that affect the activity coefficient, y_{H^+}, are the temperature, the ionic strength (I), the dielectric constant, the ion charge (Z_i), the size of the ion, and the density of the solvent (d). All these items are characteristics of the solution which relate the activity to the concentration by two main criteria. First, there is the salt effect designated as $y_{H^+}^x$. It can be approximated for the hydrogen ion in water by the expression

$$\lg y_{H^+}^x = \frac{-0.5 I^{\frac{1}{2}}}{1 + 3 I^{\frac{1}{2}}} \qquad (1.4)$$

where I is the ionic strength which is defined as one half the sum of molality in moles per kilogram, C_m, times the square of the charge of the ionic species:

$$I = \tfrac{1}{2} \sum C_m Z_i^2. \qquad (1.5)$$

(More detailed calculations of this salt effect can be found by consulting thermodynamics textbooks, in particular sections entitled Debye–Hückel equation.) The other factors listed above can be used to define these equations. Examples of the salt effect are shown in Figure 1.1. The salt effect

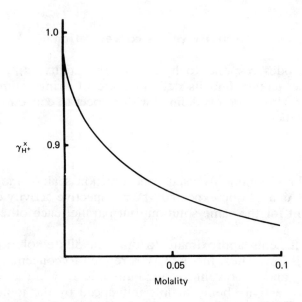

Figure 1.1 An example of an approximate salt effect

dictates that acid or base solutions having the same hydrogen-ion content have different pH values if the ionic strength of the sample varies.

Secondly, there is a medium effect designated as $y_{H^+}^m$. This effect reflects the influence that the *solvent* has on the hydrogen-ion activity. It relates the electrostatic and chemical interactions between the ion and the solvent, of which the primary and overriding interaction is solvation. Transference between different solvents can produce medium effects of several orders of magnitude (Marcus 1980).

The pH scale

In order to quantify the relative acidity or basicity of solutions it is useful to define scales of pH—such scales are relative rather than absolute, and have as their main justification the fact that they are convenient means of describing and comparing solutions.

First, we must recall that the ionization constant of water, $K_W = a_{H^+} \cdot a_{OH^-} = 10^{-14}$ mol² dm⁻⁶ for pure water at 25 °C, i.e. $[H^+] = [OH^-] = 10^{-7}$ mol dm⁻³. This is called a neutral solution, in that it has a surplus neither of the acid protons nor of the alkaline hydroxide ions (compared with the hydroxide and proton concentrations, respectively).

If pH = $-\lg a_{H^+}$, this neutral solution clearly has a pH = 7.0. Other examples are: 1 mol dm⁻³ a_{H^+} has pH = 0; 1 mol dm⁻³ a_{OH^-} has pH = 14; 10 mol dm⁻³ a_{H^+} has pH = −1; etc. However, bearing in mind our mention of temperature, salt and medium effects on previous pages, one soon realizes that these pH values cannot be stated precisely because, for example, at

Introduction

20 °C, $-\lg K_W = 14.2$ so that neutrality is now pH = 7.1. Similarly, the mineral acid or alkali used to obtain 1 mol dm^{-3} solutions contributes considerably to both the salt and the medium effects and alters activity coefficients. Thus, 10^{-2} mol dm^{-3} HCl, if the salt effect did not exist, has pH = 2.0 but since the activity coefficient, $f_{H^+}^x$, is in reality = 0.915, pH = 2.04. Further, if 90 mmol dm^{-3} potassium chloride is present in this acid, $f_{H^+}^x = 0.829$ and pH = 2.08 (and we could confuse the reader further by considering medium effects and temperature variations) (Westcott 1978).

pH measurements

The basic system is comprised of (i) a glass electrode that responds to several potentials but predominantly that of the hydrogen-ion activity, (ii) a reference electrode that is intended to be the constant potential against which all other potentials are measured, (iii) a solution under test into which (i) and (ii) dip (sometimes the reference electrode is connected via an n-shaped salt bridge to this solution) and (iv) a pH/millivoltmeter which monitors the potential difference across these two electrodes (Figure 1.2).

The various potentials which contribute to the overall potential difference monitored by the meter are shown in Figure 1.3, where E_r is the standard potential of the reference electrode, E_g is the glass-electrode potential for a given solution, $E_r - E_g$ is called the observed potential difference and is denoted as E in this text. In practice, glass electrodes are usually filled with a buffer solution which produces an internal potential equal to the potential of a saturated calomel reference electrode (E_1). Thus, $E = 0$ when the sample is

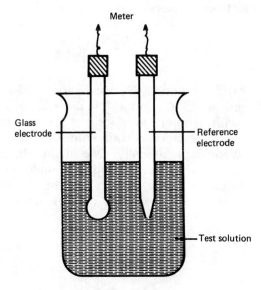

Figure 1.2 Diagrammatic potentiometric set-up

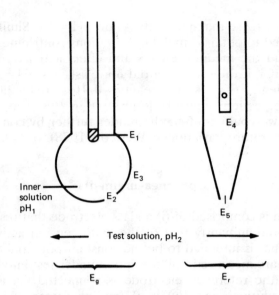

Figure 1.3 Potentials thought to be present in an electrode pair. These are discussed fully in Chapter 2

at pH = 7; for acid solutions, E is positive millivolts and for alkaline solutions E is negative millivolts.

The pH-meter, in principle, merely measures values of E and converts them into pH values. However, there are many problems to be overcome; the glass-electrode bulb has a high resistance (about 100 MΩ) and so the current that can pass through the system is extremely low. Thus, a meter which has a low bias current or a high input impedance is essential.

The theoretical electrode behaviour is

$$E = E^\ominus - \frac{2.303RT}{F} \lg a_{H^+}. \qquad (1.6)$$

$2.303RT/F$ is called the theoretical Nernst factor or slope and is equal to 59.157 mV per pH unit at 25 °C, and $E^\ominus$ is the standard potential.

The slope varies with temperature, ionic strength and liquid junction potentials. Many modern meters permit some of these variables to be 'tared-out' or allowed for *approximately*. One of the objectives of this book is to acquaint the reader with the assumptions underlying those ever-so-reassuring 'temperature' or 'slope' knobs on the meter and to present theories taking the reader as closely as possible to the absolute values measurable.

pH standards

The concept of a pH value is virtually unique compared to all other parameters commonly used in physical chemistry because pH has been

Introduction

defined in terms of hydrogen-ion activity—an unmeasurable term, since single-ion activities are indeterminate. Indeterminacy leads to controversy, in that we humans have to make assumptions in order to define a pH scale. The choice of such assumptions is somewhat arbitrary and so disagreements between scientists in different countries can occur. Thus, a pH measurement is definitely not just a matter of switching on a pH-meter, plunging the electrodes into the test solution, and taking the meter reading. Rather, there is a wide range of theoretical features to be considered, a selection of reference materials to be contemplated, and a sizeable number of parameters to be held as constant as possible whilst other variables are determined.

All pH users use *operational* definitions for the pH scale, i.e. the exact experimental steps taken in order to standardize solutions and metering devices are laid down as an operational method or recipe. Such methods refer to the platinum–H_2 electrode immersed in a solution connected via a salt bridge with a reference electrode (usually of the mercury–calomel variety). Based on this origin, standard buffer solutions of very pure chemicals are used to standardize the cell voltage at certain figures which are said to correspond to their respective equivalent pH values.

Controversy can arise concerning the choice of the exact number of standard buffer solutions which are used in order to define the pH scale (Covington 1980, 1981). In theory, providing that the activity coefficients and liquid-junction potentials of the system remain constant, the measured cell emf ought to vary in a linear manner with pH. The slope of this line should be Nernstian, i.e., 50 mV per pH unit at 25 °C. Thus, whether one measures the pH of a test solution by differences from just one pH buffer solution reference point, or by differences from several buffers, the answers ought to be equivalent. The UK uses this single-buffer approach which assumes that a potassium hydrogen phthalate solution (50.0 mmol kg^{-1} aqueous solution) at 25 °C has a pH value of 4.005. (This has recently been corrected from the older figure of 4.008.) This single primary standard scale gives a pH value with reference to one buffer solution as measured from the voltage of cells without a liquid-junction potential assuming only that the convention for ionic activity coefficients is true for this one single solution. All other pH values for any number of additional secondary standard buffer solutions are measured using this same operational cell but for these secondary buffers a carefully formed liquid junction is present. It ought to be noted that no assumptions are made herein that ionic activities for different salts or buffers at different concentrations can be interrelated and that liquid-junction potentials for different secondary solutions need to be accurately defined.

The multistandard scale, widely used in the United States and Germany, assigns pH values to five or even more buffer solutions from voltage measurements of cells without liquid junctions but assumes that the convention for ionic activity coefficients is valid for all of the buffer solutions used. These solutions are then employed to standardize operational cells. Supporters of the multistandard scale agree that the single-standard scale is attractive but it suffers from different buffered solutions having different liquid-junction potentials. Opponents suggest that the multistandard scale suffers from over-definition and sometimes leads to inconsistencies because of the additional assumptions made (compared with the single-standard scale) that the liquid-junction potentials between the buffer solutions and the salt-bridge solution are constant. The multistandard scale is known as the NBS scale (named after the US National Bureau of Standards). Different countries use differing numbers of multistandard reference materials, ranging from three in France up to nine in the Federal Republic of Germany. An excellent critique of the two approaches to buffer standards has recently been published (Bates 1981).

There is far more that unites than divides the proponents and opponents of these two scales; and such slight academic changes in pH scale definitions will have very little effect upon pH measurements in industry and medicine, which are usually satisfactory with a level of precision of ±0.05 of pH. Both sides agree that only a very small percentage of experiments require an exact interpretation of the pH number obtained. Further, whether the metering devices are calibrated with respect to a primary standard or through a secondary standard is of little consequence in real life as far larger inconsistencies arise from incorrect weighings, impure chemicals, incorrect techniques or misunderstood thermodynamic relationships! In addition, those of us who try to use pH measurements in order to establish hydrogen-ion activities or concentrations in industrial, biological, or maritime fluids, have to live with the additional problem that these solutions are of high ionic strength, and so Debye–Hückel type calculations of activity coefficients are very unreliable at such concentrations. It is important that scientific effort be applied, not to comparing different pH scales, but rather to turning pH measurements into $[H^+]$ values which can yield a rich harvest of formation constants, analytical data, and speciation and bioavailability breakthroughs.

Millivolt approaches for hydrogen-ion concentrations

Glass-electrode metrology has developed rapidly over the last three decades. From the first half of this century, when electrode voltages were converted directly into pH readings based upon rather more faith than theory concerning the exact relationship between glass electrode potential and pH difference with respect to a buffer, there has now emerged the era of millivolt measurements. Chapter 4 will describe why an ideal response is not realized in practice. The non-ideality is seen as a ΔmV value. Averaging out this non-ideality becomes a lot more difficult if the emf values are corrected onto a logarithmic scale, i.e. into pH readings.

Electrode millivolt calibrations

Calibrating electrodes in terms of concentrations rather than activities is easier to achieve from millivolts (May et al. 1982). Many of the non-linearity features arise when one calibrates an electrode pair in one titration and assumes the calibration holds for the test/research titrations. The attendant difficulty is overcome in millivolt work, and modern computer programs permit internal calibrations of the electrodes to be performed in test/research solutions.

Optimization techniques

One of the experimental techniques used to improve the precision of a measured parameter is to record the data many times under slightly differing conditions, e.g. to set up a series of cells in which one concentration is varied stepwise. A titration is a convenient means of doing this. Far more data points than are strictly required are accumulated; and then parameters calculated from such data can be optimized by a least-squares approach.

In titrations of weak acids or bases [H^+] may, in principle, be calculated at various points in the titration from the protonation constants, *provided they are accurately known*. Very precise calibrations can be made in this way but they are of somewhat limited use. In fact, the object of many titrations is to measure the protonation constants. Under such circumstances, a sizeable number of parameters, such as pK_W or a stability constant may be optimized, and even parameters which intrinsically define the system, e.g. electrode slope, $E^\ominus$, etc. Computer programs we shall be describing in this context are MAGEC and ACBA (see later). Straightforward algebraic solutions are very rarely available and so general optimization techniques are employed (May et al. 1982).

Large computer programs such as those mentioned successively estimate, check, and improve values for all unknown parameters on a cyclical basis until all are obtained as 'best' constants to within the precision of the experiment.

Formation constants

The amount of reaction products produced at equilibrium is reflected in the 'stability', 'equilibrium' or 'formation' constant (the last term is used throughout this book). When many competing reactions occur in the same vessel, each one is still governed by its formation constant. The eventual point of balance between all these equilibria is determined by, and may be calculated from, a knowledge of (i) all the relevant formation constants and (ii) all the total concentrations of the chemicals present (Dyrssen et al. 1968; Rossotti 1978).

The exact chemical composition (e.g. a complex of two ligands and one metal ion) of each chemical species present in a complicated solution at equilibrium and their respective concentrations is called the speciation. Speciation knowledge is of great value in all branches of chemistry. Unfortunately, when one tries to determine speciation by analysis one often upsets the equilibrium and obtains an unrepresentative answer. Fortunately, the correct answer may be obtained by using the 'model' calculation mentioned at the end of the previous paragraph.

Speciation, bioavailability, and the scheme of this book

Industrial companies market 'masses' or 'concentrations', not 'activities'; food and drugs are either bioavailable and can be assimilated from the intestinal tract, or inert, depending upon the nature and charge on the species present; so too in geochemistry etc. It is thus axiomatic that speciation knowledge is of fundamental importance to the understanding of such mechanisms.

As we have just outlined, the best approach to reliable speciation information is

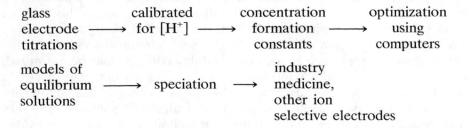

The theories underlying these topics are covered in depth in the following chapters.

2

Design and maintenance of electrodes

Reliable electrode-potential measurements require well-designed, carefully constructed electrodes. This chapter describes desirable design features, although not in great detail, since today electrodes are purchased commercially, rather than made in the laboratory. Manufacturers' catalogues list electrodes suitable for almost any task, from those with long immersion lengths of two metres, to those used in biomedical applications such as determining the pH in the stomach. Some of these are illustrated in the photograph (Figure 2.1) and listed at the end of the chapter.

Construction

The potential of a glass electrode in millivolts is composed of three components (Figure 1.3):

E_1, the potential of an internal reference electrode usually of the silver–silver chloride type;

E_2, the potential developed across the glass membrane separating an inner solution of pH_1 from that of the test solution of pH_2. This potential has the form,

$$E_2 = \frac{2.303 RTs}{F \times 100} (pH_2 - pH_1)$$

where R is the gas constant, F is the Faraday constant, T is the absolute temperature, and s is the sensitivity of the electrode response expressed as a percentage.

E_3, which is the asymmetry potential and is the residual potential which can still be detected across the membrane when $pH_2 = pH_1$.

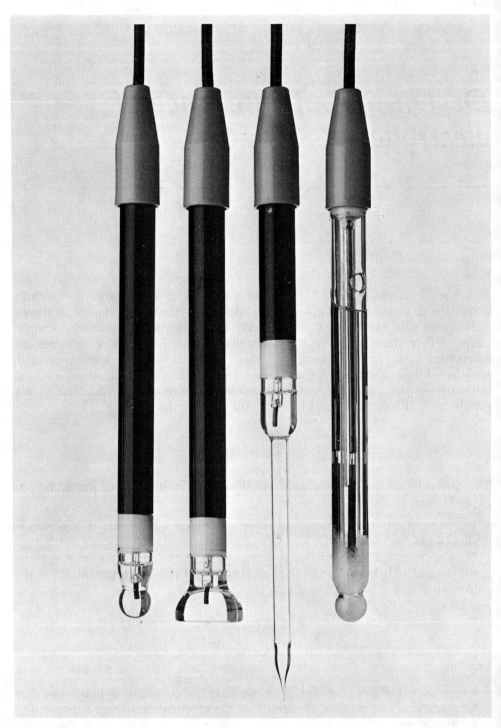

Figure 2.1 Photograph of three different types of glass electrode (regular, flat and spear) and a combination glass-reference electrode (Radiometer A/S, Copenhagen)

Design and maintenance of electrodes

The asymmetry potential is dependent to some small extent on the medium but mainly on temperature and time and it is this variation with time which is one of the reasons why a permanent electrode calibration curve cannot be determined (Linder et al. 1983).

In constructing a glass electrode a thin glass membrane usually in the shape of a bulb is sealed onto an insulated pH insensitive glass shaft. The inside of the bulb is filled with a solution of high buffer capacity towards both chloride, if the internal reference is of the silver–silver chloride type, and hydrogen ions. This results in a constant internal reference potential (E_1) and a constant hydrogen-ion concentration on the inside of the membrane (pH_1). Into this solution dips the silver–silver chloride reference electrode which is then connected via a shielded lead to the pH-meter.

Although the mechanism by means of which the glass electrode potential arises is not well understood, there is a strong possibility that hydrogen-ion-concentration-dependent ion exchange occurs on anionic sites, mainly provided by the silicon dioxide of the glass, on both sides of the glass membrane, and because the hydrogen-ion concentration is different on the two sides a potential difference occurs. For the device to serve as a useful electrode this potential must be stable and reproducible and there must be conductance across the membrane. In addition, the response of the electrode to other ions such as sodium and lithium must be constant and minimal.

In general, the design, composition and thickness of electrode glasses, are a compromise between robustness, conductance and sensitivity.

Glass composition

When a liquid is rapidly cooled it either crystallizes or forms a glass. Most compounds which are solid in the range 0–100 °C (the electrode is to be used in water solutions) form crystals and this limits the choice of compounds which can be commercially exploited for forming glasses to oxides and salts such as silicates or others containing oxyanions. In the main, arsenic(III) oxide (As_2O_3), boron(III) oxide (B_2O_3), germanium(IV) oxide (GeO_2), phosphorus(V) oxide (P_2O_5) and silicon(IV) oxide (SiO_2) are used either pure or, more recently, in combination. Alkali oxides and alumina which themselves do not give glasses may be added for special reasons. Commercial glass electrodes all contain a minimum of 50% SiO_2 for reasons of chemical durability.

A typical pH range 0–14 electrode bulb glass is:

SiO_2, 63%; Li_2O, 28%; Cs_2O, 2%; BaO, 5% and La_2O_3, 2%.

Electrodes to be used in the pH range 0–11 have the caesium and lanthanum oxides omitted giving rise to a product which has lower resistance than the full-range electrode. These narrow range electrodes are suitable for use in non-aqueous media and at low temperatures.

Robustness

Ideally, a glass electrode ought to be capable of giving many months of service which would involve knocks, high temperatures, and attacks by corrosive solutions. Modern technology has come a long way towards achieving these objectives.

First, the production of sensitive meters capable of detecting very low electrode currents has permitted the use of thicker, more durable glass bulbs, with higher membrane resistance, without a reduction in sensitivity.

Secondly, the glass resistance change with ageing is now known to be glass-composition dependent. For example, slightly lowering the lithium content can markedly improve ageing characteristics. Some glasses are more soluble than others. This glass dissolution becomes a problem at alkaline pH values because metasilicic acid reacts with alkali, the process commencing at pH values as low as 8.5:

$$H_2SiO_3 + NaOH \rightarrow NaHSiO_3 + H_2O.$$

Bates (1973) discussed the pH response and relationship to solubility for a wide range of glasses.

Modern scientific research is fortunate in that half a century of knowledge and improvements along the lines illustrated have been incorporated into modern commercially available electrodes.

Filling solutions

Both the glass and reference electrodes require filling solutions; over the decades potassium chloride has evolved as the usual first choice because it seems to satisfy most of the criteria, especially that of the diffusion rates of anion and cation being equal. There are, however, instances when potassium chloride has to be avoided; for example, when the test solution contains perchlorate. Leakage of potassium ions from the reference electrode can produce a precipitate of potassium perchlorate which can then block the liquid junction. Similarly, chloride leakage into a solution in which chloride is being assayed is clearly undesirable. Thus, filling-solution electrolytes for reference electrodes ought always to be considered from the viewpoint of chemical reaction with the sample, contamination of the sample, and solubility in solvents present.

Reference electrodes

Single-electrode potentials are analogous to single-ion activities in that they are not experimentally measurable in absolute terms. Consequently, in practice, a second potential is needed to permit a potential difference to be determined. pH measurement using an electrochemical cell, therefore, requires two electrodes, one being the working hydrogen-ion–concentration-sensitive electrode and the other a hydrogen-ion–concentration-insensitive electrode (Figure 1.3).

A reference electrode embodies an internal electrode such as a calomel or silver–silver chloride cell (potential = E_4) and an electrolytic filling solution, the latter being contained in a glass or polymer salt bridge which surrounds the internal electrode and makes electrical contact with the test solution through the liquid junction (potential = E_5).

The internal element of a calomel electrode is a grey cylindrical tablet which has mercury at the top and solid mercurous chloride compacted into the tablet below the metal. Provided that the filling solution is saturated potassium chloride, the mercury–mercurous chloride half-cell provides a potential of 244 mV versus the normal hydrogen electrode at 25 °C. A standard silver–silver chloride reference electrode provides a voltage of 199 mV provided a filling solution saturated with both potassium chloride and silver chloride is used. The silver–silver chloride internal element is a metallic silver wire coated with silver chloride.

The silver–silver chloride electrode is preferable to the calomel electrode at temperatures greater than 50 °C because at high temperatures calomel is subject to disproportionation:

$$Hg_2Cl_2 \rightleftharpoons Hg + HgCl_2.$$

This reaction is slow at low temperatures but above 70 °C the stability of the calomel potential is markedly affected. At lower temperatures the slow attainment of the above equilibrium contributes to the appreciable temperature hysteresis exhibited by the calomel electrode.

When measurements are being made the filling hole of the electrode must be left open to the atmosphere, otherwise the flow of liquid through the liquid junction will stop. A slow leakage through the junction is needed in order to maintain electrical contact with the surrounding solution.

The liquid junction is an important feature of the reference electrode. Junctions are made from various materials—asbestos, quartz fibres, ceramic plugs and even carborundum frits or ground-glass sleeves have been used. They differ in terms of the liquid-junction potential set up and the rate at which the filling solution leaks through the junction. All junctions ought to be kept wet at all times to prevent clogging. Also, the head of the solution forming the filling solution ought to be greater than that of the sample solution or else the sample will be forced into the inside of the reference electrode, and this can lead to unstable potentials. Whenever the crystals at the bottom of the electrode become hard-packed or glassy in appearance, the electrode is in danger of junction clogging, and it should be shaken gently and more filling inserted if necessary. Further advice on choice of junction and the treatment of clogged junctions is given in Westcott's book (Westcott 1978), and by Linnet (Linnet 1970).

Combination electrodes

The combination of glass and reference electrode into a single probe produces a combination electrode. In general, the behaviour described in the previous sections for the separate glass and reference electrodes also applies to combination electrodes. The reference potential in a combination electrode is usually the silver–silver chloride system, as this is thinner and is more easily positioned than a calomel tablet. A typical combination electrode is shown in Figure 2.1.

The liquid junction can vary from a circular ceramic ring surrounding the glass bulb to a small glass sinter sealed into the side of the outer glass tube. In all instances one must aim to minimize the strain on the glass electrode bulb. The inside glass electrode relies on the outer reference-electrode solution for partial shielding and it is therefore necessary to maintain a good head of outer reference electrode filling solution in order to maximize this shielding feature.

The main advantages of combination electrodes are that they permit small volumes to be monitored, and measurements can also be made more conveniently in regions of restricted access, such as in industrial flowing pipes or meat samples. Disadvantages are that there is more interference and more risk of bulb strain. Also, the cost of replacement of a combination electrode is greater than that of replacing either part since both portions have to be discarded if the bulb should break.

Commissioning a new electrode

Before a new glass electrode or one which has been stored dry can be used, it must first be soaked in a suitable solution, so as to form a swollen protonated layer of low resistance and to stabilize the asymmetry potential. Linnet recommends the following procedure (Linnet 1970). Soak the electrode for at least 24 hours in $100\ \text{mmol}\ \text{dm}^{-3}$ hydrochloric acid at room temperature, rinse it afterwards in distilled water and then resoak for several hours in a buffer solution with pH between 4 and 8. During the first few days of use subsequently, further soaking will occur, giving rise to a decreasing response time.

Chapter 4 describes how electrodes are assessed in quantitative detail, but it is also relevant to this electrode design chapter to list the features to look for when assessing electrodes—these can be quite simple to test using a millivoltmeter and another good electrode.

Each new electrode can be measured with reference to an equivalent well-known and well-tried electrode of the same type, i.e. glass versus glass or reference versus reference. Clearly, the meter reading ought not to change as the sample solution in which both electrodes are immersed is varied. In a similar vein the sensitivity can be tested by a plot of mV versus the pH of a series of standard buffers. A straight line should be obtained with approximately the theoretical slope. This buffer-line determination should be carried out frequently during the lifetime of the electrode and when the slope falls to approximately 95% of the theoretical the electrode should either be discarded or attempts made to regenerate it, perhaps by etching. The following etching method has been suggested by Linnet (Linnet 1970). Immerse the electrode for 1 minute in a 20% ammonium bifluoride solution; then flush with water, immerse electrode in 5 mol dm^{-3} hydrochloric acid for 10–20 seconds, rinse in distilled water and store in a buffer solution (pH 4 to 8) for at least 1 hour.

In an emergency, the electrode may be dipped into a solution of approximately 1 mol dm^{-3} hydrogen fluoride for approximately one minute, as suggested by Schwarzenbach (private communication, 1975). However, it must be emphasized that etching shortens the life of electrodes considerably and should be used as a last resort.

Generally, an electrode can be considered in good order if:

(1) The sensitivity as determined from buffers is better than 95%.
(2) The response is fast (<1 minute).
(3) There is a stable response during stirring.

Electrode storage

For storage periods of long duration the glass electrode should be stored dry. The reference electrode should also be stored dry and sealed, or with its tip immersed in the salt-bridge solution. Before taking the glass electrode into use again it must, of course, be soaked.

For short-term storage, i.e. between titrations, or overnight, it is important to keep the surface layer on the glass intact. Thus, it can be kept in a solution of approximately pH 4 or in distilled water. Electrodes that are not customarily taken to the acid end of the pH scale should be dipped occasionally in mildly acid solution in order to decrease the response time.

Reference electrodes should not be stored in distilled water or weak solutions as this can lead to dilution of the filling solution. Whenever the electrode is stored for more than an hour or so, it is important to ensure that it is placed in a vessel in which the electrode stem and vessel are closely joined—by using B14 ground-glass joints, for example. This prevents evaporation of the storage solution and thus prevents the formation of a crust of salt on the tip of the electrode. Reference electrodes such as the calomel electrode which have a slow temperature response are best stored by having their storage solution vessels fastened to the walls of a thermostat bath kept at the temperature at which the potential measurements are being made.

Cleaning and transferring electrodes

Electrodes are invariably washed as part of the transference procedure from one solution to another. This is conveniently done by giving them three large drenchings with deionized water and then blotting them dry with tissues. Electrodes should be blotted dry, and not wiped, in order to minimize the build up of static electricity. Build-up charge on the electrode may only be slowly dissipated in the sample solution, thus leading to a longer response time. Rinsing with distilled water and then with some of the solution to be tested can be used when the absolute volume of the test solution is not important. Obviously, this method cannot be used when a titration is to be performed. Water used in all electrode manipulations and solutions should, if reasonable precision and accuracy of measurement is required, be of good quality and glass-distilled.

When the electrodes are immersed in the sample solution one has the choice of stirring by passing an inert gas through the solution, by using a magnetic follower, or by merely shaking the sample. In most laboratories magnetic stirring, sometimes combined with an inert gas atmosphere to prevent the entry into the system of gases such as oxygen and carbon dioxide, is used. Stirring allows a homogeneous solution to be obtained giving a fairly fast response time but all calibration procedures must be carried out at the same stirring rates. Also, care must be taken to ensure that the motor driving the stirrer does not introduce heat into the solution, and that the magnetic follower does not bump when rotating and thereby damage the electrodes.

Meters

The high resistance of the glass electrode requires that the input impedance of the meter used to measure the potential of the electrode train be high ($>10^{11}$ ohm). Moreover, the meter must draw only a small current ($<10^{-11}$ amp) to avoid polarization of the reference electrode and to avoid the occurrence of an undesirable voltage drop across the glass membrane. Also, the electrical zero to which the electrode train potential is being referred must be exceedingly stable.

To meet these requirements the meters used all employ electronic circuitry. The earlier ones used radio valves and were of the analogue type in which the potential was read by means of a needle moving along a calibrated scale. Today all pH meters depend on solid-state components and analogue read-out is giving way to digital. Potentials can be measured to 0.1 mV as a matter of course and to 0.01 mV with very little extra effort in a thermostatted room. The advent of microprocessors has led to automatic titrators—in which the potential at each point is measured, stored in memory, plotted and endpoints automatically calculated—becoming almost routine features in research laboratories.

Glass electrodes for special purposes

Below is a brief list of some electrodes which have been designed for special

Design and maintenance of electrodes

purposes (Crow 1979):
- (i) Electrodes which can withstand high pressures and high temperatures
- (ii) Electrodes which require only small sample volumes, e.g. 20 μ dm^3
- (iii) Electrodes with long immersion lengths, e.g. two metres
- (iv) Very small electrodes for biomedical application, e.g. stomach pH
- (v) Electrodes with flat sensing heads for skin, paper and textiles
- (vi) Spear-headed electrodes for cheese, meat, and soil
- (vii) Robust electrodes, plastic coated, gel-filled with epoxy coating.

3

Determination of pH values

Introduction

Before considering the procedural steps required to measure the pH of a given solution we need, first of all, to outline the background theory concerned with the definition of the property, pH, and the establishment of the pH scale.

Two main difficulties surrounding the measurement of pH as defined by Sørensen, namely,

$$\mathrm{pH} = -\lg[\mathrm{H}^+] \tag{3.1}$$

may be summarized as in (i) and (ii) below (Sørensen 1909):

(i) Physical measurements tend to respond to hydrogen activities, a_{H^+}, rather than to hydrogen-ion concentrations, $[\mathrm{H}^+]$. The meaning of activity has been mentioned in Chapter 1; and the subject is enlarged upon in the next section of this chapter. Sørensen and Linderstrøm-Lang (Sørensen and Linderstrøm-Lang 1924; Sørensen *et al.* 1927) dealt with this difficulty by modifying the definition of pH to

$$\mathrm{pH} = -\lg a_{\mathrm{H}^+} \tag{3.2}$$

but subsequently certain additional difficulties became recognized. These, and the manner of dealing with them, are discussed later in this chapter.

(ii) The proton in water is highly hydrated, consisting mainly of the species, $\mathrm{H_3O^+}$, but also to some extent of $\mathrm{H_5O_2^+}$, $\mathrm{H_7O_3^+}$, $\mathrm{H_9O_4^+}$ and other hydrates. This difficulty is readily resolved in the present context simply by using the symbol $[\mathrm{H}^+]$ to represent the sum of the concentrations of all forms of hydrated protons. Likewise, by a_{H^+} is understood the sum of the activities of all forms of hydrated protons.

The definition (3.2) would be totally useful and universally applicable were it possible to measure single-ion activities. Unfortunately, this is not so and hence the definition of an additional property is required. Let us illustrate this problem.

Nowadays, the most important way of measuring pH is to determine the emf of an appropriate electrochemical cell (Bates and Guggenheim 1960); for example,

| $Pt,H_2(g)$ | experimental solution containing H^+ (and OH^-) and an anionic species, e.g. Cl^- | reference electrode (e.g. $AgCl(s)\,|\,Ag(s)$) |

The potential of the left-hand (hydrogen) electrode depends strongly on the proton concentration (or activity) of the experimental solution. That of the reference electrode, however, is independent of $[H^+]$ and a_{H^+}.

Application of the Nernst equation (Bates 1981) gives us the emf, E_{cell}, of the cell:

$$E_{cell} = E_{ref}^\ominus - k \lg a_{Cl^-} - E_{H_2}^\ominus + k \lg p_{H_2}/a_{H^+} \quad (3.3)$$

where

$E_{ref}^\ominus$ is the standard potential of the reference electrode;
$E_{H_2}^\ominus$ is the standard potential of the hydrogen electrode, $=0$ by definition;
$k \quad = RTF^{-1} \ln(10)$;
$R \quad$ = the universal gas constant;
$T \quad$ = the temperature of the cell in kelvins;
$F \quad$ = the Faraday constant;
a_{Cl^-} = the activity of chloride ions in the experimental solution;
p_{H_2} = the pressure of the hydrogen gas.

It follows that

$$E_{cell} = E_{ref}^\ominus - k \lg a_{H^+} a_{Cl^-}/p_{H_2} \quad (3.4)$$

$E_{ref}^\ominus$ is available from the literature and p_{H_2} can be measured. Hence, E_{cell} provides a measure of the product $a_{H^+}a_{Cl^-}$. Thus, the emf measurement gives us a value for the *product* of the activities of the protons and the anions in the experimental solution. This result is general: *an electrometric* (or any other type of physical) *measurement cannot give us estimates of single-ion activities, but only of activity products for the cationic and anionic species present.*

Activities and activity coefficients

Dissolved electrolytes interact with aqueous solvents in such a manner that they tend to behave as though their concentrations differ effectively from their true concentrations. A well-known example in which this non-ideal behaviour is manifested is in the van't Hoff *i* factor derived from measurements of osmotic pressure, depression of freezing points or elevation of boiling points. For a strong 1:1 electrolyte such as sodium chloride, the expected value of *i* is 2, corresponding to complete dissociation into two ionic species (one cationic and one anionic). The deviation of observed i values from the expected value (e.g. for 100 mmol kg^{-1} NaCl, $i = 1.87$) indicates that the effective concentration of the ions is lower than the true concentration, even though the electrolyte is completely dissociated.

The effective concentration of a solute is termed its activity, which is generally denoted by the symbol a. It is customary to express the activity in terms of concentration. In the following equations (3.5) we give the relationship, using three important ways of expressing concentration of the solute:

$$\left. \begin{array}{l} a_c = yc \\ a_m = \gamma m \\ a_N = fN \end{array} \right\} \quad (3.5)$$

Here, $c =$ the molar concentration (e.g. in mol dm^{-3}); $m =$ the molality (e.g. in mol solute per kg of solvent); $N =$ the mole fraction of the solute; y is called the molarity activity coefficient; γ is called the molal activity coefficient, and f is called the rational activity coefficient of the solute.

a_c, a_m and a_N refer to the activities relating to the three ways of expressing concentration. In the treatment that follows, we generally use molal concentrations but we drop the subscript m and simply use the symbol a.

The values of the activity coefficients vary with the concentration, the actual value, in particular its deviation from unity, giving a measure of the deviation of the solution from ideal behaviour. In dilute solutions, the values of y, γ and f tend to become sufficiently close for the differences to be ignored in most circumstances. Usually as the concentration approaches 0, the activity coefficient approaches 1, indicating the approach of ideal behaviour with the activity approximating closely to the concentration.

The thermodynamic definition of activity is given by equation (3.6)

$$G = G^\ominus + RT \ln a \quad (3.6)$$

where G represents the Gibbs free energy or chemical potential of the solute and $G^\ominus$ is the value of G when the solute is in a specified standard reference state.

The conventionally chosen standard state is that in which the solute has unit activity. Thus, strictly, equation (3.6) should be written

$$G = G^\ominus + RT \ln a/a_{\text{ref}}$$

with $a_{\text{ref}} = 1$. If we combine equations (3.4) and (3.5), we obtain, for example,

$$G = G^\ominus + RT \ln \gamma m.$$

As a solution becomes more dilute there is a tendency, therefore, for the

Determination of pH values

expression

$$G = G^{\ominus} + RT \ln m$$

to become more valid.

It is important to realize that activity coefficients vary not only with the concentration of the solute concerned but also with the total ionic strength of the solution, for example, if two or more electrolytes are added. The ionic strength, I, of a solution is defined by the expression

$$I = \tfrac{1}{2} \sum_i m_i z_i^2 \qquad (3.7)$$

where m_i denotes the molality and z_i the charge number (positive for a cation and negative for an anion) of the ion, i, in the solution. I, therefore, represents some sort of measure of the total concentration of ionic charge in the solution.

Another factor that influences the activity coefficient of an ionic species is the nature of the solvent. The latter can usually be expressed in terms of the relative permittivity (dielectric constant). This factor needs to be taken into account when dealing with mixed solvent systems, e.g. alcohol/water, dioxane/water.

Activity coefficients of electrolytes may be measured experimentally, for example, by means of emf measurements of galvanic cells, solubility measurements, and freezing-point depressions (Levitt 1973; James and Pritchard 1974). But, as implied by the example of cell (3.3) and equation (3.4), all such measurements on a given electrolyte solution yield an activity coefficient which is a property of both the cationic and anionic species present. This property is termed a *mean ionic activity coefficient* and may be denoted by $y_{\pm}$, $\gamma_{\pm}$, or $f_{\pm}$, analogous to the definitions of equation (3.5), and corresponding to the three main types of concentration unit. Alternatively, mean ionic activity coefficients may be calculated in most circumstances, to a reasonable approximation, by means of the extended Debye–Hückel equation (Glasstone and Lewis 1960) (equation (3.8)) or one of the reduced forms thereof. For an electrolyte, $M_x L_y$,

$$\lg \gamma_{\pm} = -\frac{A z_+ z_- I^{\frac{1}{2}}}{1 + B \aa I^{\frac{1}{2}}} + cI \qquad (3.8)$$

where A and B are constants characteristic of the solvent ($A = 0.509$; $B = 0.328$ for water at 25 °C)

z_+ is the charge on M^{z+}
z_- is the charge on L^{z-}
$\aa$ is an ionic size parameter
c is an empirical constant, characteristic of the electrolyte.

Equation (3.8) yields reasonably reliable mean ionic activity coefficients for solutions of total ionic strength up to about 1 mol kg^{-1}. For ionic strength values below about 100 mmol kg^{-1}, the product cI becomes negligible and equation (3.8) reduces to what is known as the Debye–Hückel equation. The product, $Bå$, becomes negligible by comparison with unity when the ionic strength of the solution is reduced to below about 10 mmol kg^{-1}, in which case the equation reduces to the Debye–Hückel limiting law.

Although only mean ionic activity coefficients may be obtained from experimental measurements, there is a basic desire to attempt a division into individual ionic contributions. This is especially important in trying to use the Sørensen and Linderstrøm-Lang definition of pH, equation (3.2). Thus, for a solution of a given electrolyte, M_xL_y, we regard the cations and anions as having molal activity coefficients γ_+ and γ_-, respectively. It is customary to express the mean molal ionic activity coefficient in terms of the individual ionic activity coefficients as in equation (3.9):

$$\gamma_\pm = (\gamma_+ \gamma_-)^{1/(x+y)}. \tag{3.9}$$

Similar relationships hold for the molarity and rational activity coefficients.

Whereas single-ion activity coefficients are inaccessible directly from experimental measurements, they may be calculated theoretically without any difficulty, provided the total ionic strength is below about 10 mmol kg^{-1}. For this purpose use can be made of equation (3.10), which is a form of the Debye–Hückel limiting law applicable to single ionic species (cf. the reduced form of equation (3.8) which yields mean ionic activity coefficients):

$$\lg \gamma_i = -Az_i^2 I_i^{\frac{1}{2}}. \tag{3.10}$$

γ_i is the molal ionic activity coefficient of the ion, i, in the solution. When the ionic strength lies between 10 and 100 mmol kg^{-1}, equation (3.11) may be used:

$$\lg \gamma_i = -\frac{Az^2 I^{\frac{1}{2}}}{1 + Bå I^{\frac{1}{2}}}. \tag{3.11}$$

Here, a difficulty arises, however, because the size parameter, $å$, is a property of the electrolyte, not of a single ionic species; its value, therefore, is affected by the nature of the counter-ion. To overcome the difficulty, there is a need to replace $å$ by some distance parameter, $å_i$, which is specifically a property of ion, i. In finding a solution to this problem, Kielland (1937) adopted a convention introduced by MacInnes (1919) that the cation and anion in aqueous potassium chloride solutions of any concentration have equal activity coefficients. The MacInnes convention is rationalized by virtue of the similar electronic structures and mobilities of potassium and chloride ions (Bates 1973). Kielland (1937) also made use of a convention introduced by Guggenheim (1930) which, when applied to potassium chloride solutions, yields equation (3.12):

$$\gamma_{K^+} = \gamma_{Cl^-} = \gamma_{\pm KCl}. \tag{3.12}$$

Determination of pH values

The implication of considering equations (3.11) and (3.12) jointly is

$$\text{å}_{K^+} = \text{å}_{Cl^-} = \text{å}_{KCl}. \tag{3.13}$$

By applying experimental values of the mean ionic activity coefficient of potassium chloride to equations (3.11), (3.12) and (3.13), a value for å_{Cl^-} may be obtained. Kielland (1937) adopted the latter as a reference together with the assumption that γ_i for a given ionic species is determined solely by the total ionic strength and is independent of the composition of an electrolyte solution. For example, he assumed γ_{Cl^-} to have the same value in 50 mmol kg^{-1} potassium chloride as in 50 mmol kg^{-1} sodium chloride.

With this approach, Kielland built up a set of å_i values (extrapolated to $I = 0$) for a large number of ions in aqueous solution. The Kielland å_i values may thus be used to calculate γ_i values for single ionic species from equation (3.11), applicable to ionic strengths up to 100 mmol kg^{-1}. The general uncertainty is about 5 per cent, which is sufficiently small to let these values be of significant use for many purposes. In attempting to calculate γ_i for ionic strengths greater than 100 mmol kg^{-1}, the difficulties magnify considerably with a consequent escalation of the uncertainty in the values obtained.

The basic problem here is to find suitable values for the parameter c in equation (3.8), or its analogue applicable to single ionic species. Some ways of obtaining c values are discussed by Linder and Murray (1982) but these need not concern us in the present text provided we limit our attention to electrolyte solutions of ionic strengths lower than 100 mmol kg^{-1}. It must be emphasized that, except for very dilute solutions, single-ion activity coefficients which are calculated as described above are *relative* values based on an assumed value for å_{Cl^-}, which is taken as a reference value; this involves the additional assumptions outlined above. There is, in fact, no rigorous way of obtaining a single-ion activity coefficient.

Operational definition of pH

We have seen that the measured emf of the cell (3.3) yields a value of the activity product, $a_{H^+} a_{Cl^-}$, for the experimental solution. By invoking equation (3.9), the activity product value can be converted to

$$a_{H^+} m_{Cl^-} \gamma_{Cl^-}.$$

m_{Cl^-} being separately measurable. The measured emf of the cell can, therefore, be made to yield a value of $a_{H^+} \gamma_{Cl^-}$, and hence of $-\lg a_{H^+} \gamma_{Cl^-}$, or pH $-\lg \gamma_{Cl^-}$. Thus, γ_{Cl^-} is the 'additional quantity' referred to in the introduction to this chapter that needs to be defined in order to obtain a pH value, by combining it with a measured emf. The difficulties surrounding the evaluation of single-ion activity coefficients, such as γ_{Cl^-}, are discussed in the previous section. To overcome this and certain other difficulties, the relevant international and national bodies such as the International Union of Pure and Applied Chemistry (IUPAC) and the US National Bureau of Standards (NBS), have introduced an operational definition of pH; i.e. pH is defined in terms of the operation or method used to measure it. The definition makes use of the operational cell (Covington 1981), (3.14), or minor variants of it:

$$\text{Hg} | \text{Hg}_2\text{Cl}_2 | \text{KCl} (\not< 3.5 \text{ mol dm}^{-3}) \left\| \begin{matrix} \text{solution} \\ \text{S or X} \end{matrix} \right| \text{H}_2 \Big| \text{Pt.} \quad (3.14)$$

First, the emf, E_S, is measured when the cell contains a solution, S, of assigned pH value, pH(S). Subsequently, the solution, S, is replaced by the unknown (test) solution, X, and the corresponding emf, E_X, is measured. The pH of solution X is then given by equation (3.15):

$$\text{pH}(X) = \text{pH}(S) - (E_X - E_S)/k - (E_{J(X)} - E_{J(S)})/k \quad (3.15)$$

where $k = RTF^{-1} \ln(10)$ (see equation (3.4)).

$E_{J(X)}$ and $E_{J(S)}$ are the liquid-junction potentials arising in the cell when solutions X and S, respectively, are present. The term $E_{J(X)} - E_{J(S)}$ is referred to as the residual liquid-junction potential and in the operational definition of pH(X) is *assumed* to be 0. The latter represents a more reasonable approximation, the more similar X and S are in composition and concentration.

Thus, the operational definition of the pH of a solution X depends on the assignment of pH(S) values to a selected standard reference solution. This aspect is covered in the next section. As Covington (1981) emphasizes, the intention behind the operational definition of pH is that it should yield numbers that are reproducible from place to place and do not vary with time.

Definition of the pH scale

In order to obtain a pH(S) value for a suitably chosen standard reference solution, S, a cell of the type (3.16), without a liquid junction, is employed (Bates, 1981):

$$\text{Pt} | \text{H}_2(1 \text{ atm}) | \text{solution S, Cl}^-(m_{\text{Cl}^-}) | \text{AgCl} | \text{Ag.} \quad (3.16)$$

The emf can be represented by equation (3.17), a form of the Nernst equation:

$$E = E^{\ominus} - k \lg m_{\text{H}^+} \gamma_{\text{H}^+} \gamma_{\text{Cl}^-}. \quad (3.17)$$

The cell consists of the standard reference solution S to which chloride has been added.

$m_{\text{H}^+}, m_{\text{Cl}^-}, \gamma_{\text{H}^+}, \gamma_{\text{Cl}^-}$ are the molalities and single-ion activity coefficients of hydrogen and chloride ions, respectively, in the cell solution;

k, as in equation (3.4), represents the Nernstian slope factor, $RTF^{-1} \ln(10)$;

Determination of pH values

$E^\ominus$ is the value of E when the cell solution consists of hydrochloric acid solution with an activity of unity and no additional chloride.

Equation (3.17) may be rearranged to yield (3.18):

$$(E - E^\ominus)k^{-1} + \lg m_{Cl^-} = -\lg m_{H^+}\gamma_{H^+}\gamma_{Cl^-}. \quad (3.18)$$

In order to determine pH(S), a series of measurements of the cell emf is made corresponding to various concentrations of added chloride. Since the value of $E^\ominus$ is known (Antelman 1981), the left-hand side of equation (3.18) can be evaluated for any of the chloride concentrations concerned. The values so obtained are plotted and extrapolated to $m_{Cl^-} = 0$. The extrapolated value is equal to $-\lg m_{H^+}\gamma_{H^+}\gamma_{Cl^-}$ for solution S with the effect of added chloride thus eliminated. Let us denote this extrapolated value by

$$(-\lg m_{H^+}\gamma_{H^+}\gamma_{Cl^-})^0$$

or alternatively by

$$p(a_{H^+}\gamma_{Cl^-})^0.$$

The latter may be expanded as follows:

$$p(a_{H^+}\gamma_{Cl^-})^0 = (pa_{H^+})^0 - (\lg \gamma_{Cl^-})^0$$

where $(pa_H)^0$ is the value of pH(S) being sought. Hence,

$$\text{pH(S)} = (-\lg m_{H^+}\gamma_{H^+}\gamma_{Cl^-})^0 + (\lg \gamma_{Cl^-})^0. \quad (3.19)$$

The first term on the right-hand side of equation (3.19), then, is the experimentally determined quantity. In the second term, γ_{Cl^-} can be taken as the activity coefficient that a trace of chloride would have in the standard reference solution S. As discussed earlier, there would be no difficulty in evaluating $(\lg \gamma_{Cl^-})^0$ if the ionic strength of solution S were lower than about 10 mmol kg^{-1}. In practice, however, solution S or, for that matter any buffer solution, must have a sufficiently high ionic strength in order to achieve a reasonable buffer capacity. The requirement here is that the pH must not be detectably affected by traces of impurities either originally in the constituents of the solution or from the environment. In order to strike a compromise between a reasonable level of buffer capacity and tractability of γ_{Cl^-}, ionic strengths up to 100 mmol kg^{-1} are taken as appropriate. At this level of ionic strength, some of the problems pointed out already for evaluating single-ion activity coefficients have to be considered. Taking these into consideration, a proposal put forward by Bates and Guggenheim (1960) for solutions of ionic strength not exceeding 100 mmol kg^{-1} has become universally adopted, namely, that γ_{Cl^-} be *defined* according to the *convention* of equation (3.20).

$$\lg \gamma_{Cl^-} = -AI^{\frac{1}{2}}/(1 + \rho I^{\frac{1}{2}}), \quad (3.20)$$

where A is the Debye–Hückel factor (refer to equation (3.8)) and the recommended value for ρ is $1.5\,\text{mol}^{-\frac{1}{2}}\text{kg}^{\frac{1}{2}}$. This value for ρ, incidentally, implies a value of 450 pm for $\mathring{a}_{Cl^-}$. It follows from equations (3.19) and (3.20) that pH(S) is expressed in terms of a *conventional* or *hypothetical* single-ion activity coefficient for chloride. The interpretation of pH(S) by means of the Sørensen and Linderstrøm-Lang definition (equation (3.2)) is accordingly limited by how closely the conventional γ_{Cl^-} approximates to the true γ_{Cl^-} value. Although this cannot be known, it is believed that the approximation is indeed very close.

The final step in establishing the pH(S) value for a given standard reference solution is to repeat the above procedure at several different temperatures, T, and to fit the results to an expression of the form of equation (3.21), with T in Kelvins:

$$\text{pH(S)} = AT^{-1} + B + CT + DT^2. \qquad (3.21)$$

Thus, although there is some uncertainty in the interpretation of pH(S), owing to the dependence on a conventional (hypothetical) chloride activity coefficient, the above procedures ensure that measured pH(S) values are definitive numbers which are precise (to three decimal places) and reproducible, capable of forming a universal reference scale provided the composition of the standard reference solution is carefully chosen. This treatment forms the basis of defining the pH scale, but the details differ from one country to another.

In the UK, what is known as the single primary standard approach is adopted (Covington 1981). It is deemed sufficient to establish the pH(S) value for but one carefully selected standard reference solution and, at the same time, to define the slope, k, of the emf, E, versus pH line for the cell (3.16). The standard reference solution chosen is 50 mmol kg^{-1} potassium hydrogen phthalate prepared from precisely specified constituents. The value of k is specified as the theoretical Nernstian slope, namely, $RTF^{-1}\ln(10)$. The most recently measured values of pH(S) for potassium hydrogen phthalate are given by Bütikofer and Covington (1979). For example, the value for 25 °C is 4.005, which differs slightly from the previously accepted value of 4.008. In addition to the single primary standard, the British specification allows the use of any number of secondary standard reference solutions with pH values established by means of the operational cell (3.14) and application of equation (3.15). The construction of the operational cell is specified in detail with cylindrically symmetrical liquid junctions formed within 1 mm capillary tubes. The latest version of the British standard includes pH values for 16 secondary standard reference solutions (Covington 1981).

In many other countries, the multiprimary standard approach is adopted. The pH(S) value for each standard is established, as outlined above, using the cell (3.16) together with the Bates–Guggenheim convention for defining γ_{Cl^-}. The US National Bureau of Standards (Bates 1981) specifies seven primary standards, including 50 mmol kg^{-1} potassium hydrogen phthalate. Examples of the number of primary standards specified in other countries are as follows: France, 3; Romania, 5; Poland and Hungary, 7; Federal Republic of Germany, 9.

Bates (1981) points out that in theory any two of the NBS reference solutions can be used to fix the pH scale and, indeed, that each assigned standard value ought to be an equally valid estimate of the conventional pH(S). Nevertheless, slight inconsistencies in the standardization of the operational pH scale are unavoidable in view of the differences in residual liquid-junction potential when one solution replaces another. Bates shows, however, that the inconsistencies with the majority of the standard buffers amount to less than 0.01 pH unit at 25 °C. (Slightly larger inconsistencies are found at both low and high temperatures.)

Amongst the 16 secondary standards specified in Britain, 7 have identical compositions to the US primary standards, including 50 mmol kg^{-1} potassium hydrogen phthalate. There are slight discrepancies in the respective pH(S) values given by the two national bodies, however. Covington (1981) attributes these discrepancies to the following factors:

(i) The US (NBS) value for 50 mmol kg^{-1} potassium hydrogen phthalate is based on results reported by Hamer et al. (Hamer and Acree 1944; Hamer et al. 1945, 1946), which have subsequently been shown to be slightly in error (Bütikofer and Covington 1979).
(ii) Adoption of the Bates–Guggenheim convention for each standard, implying doubts about the strict validity of this convention amongst the whole range of NBS standards.

With regard to the second of these factors, Bates (1981) has examined the effects of adopting several different reasonable conventions. He shows that for ionic strengths not exceeding 100 mmol kg^{-1}, the corresponding γ_{Cl^-} values lead to pH values which agree with one another to within a few thousandths of a pH unit. He argues that the slight discrepancies are more likely to arise from differing liquid-junction potentials, particularly when the pH extends to values lower than about 3 or greater than about 10.

Much controversy reigns over whether the pH scale should be defined by the single primary standard or the multiprimary standard approach and the matter is currently under debate within the IUPAC Commissions on Electrochemistry and Electroanalytical Chemistry (IUPAC 1980; Covington et al. 1983).

Covington (1981) is of the opinion that the multiprimary standard approach constitutes over-definition and must inevitably lead to inconsistencies as discussed above. He argues that adoption of the single primary standard pH scale represents a simplification of existing procedures and is the only correct one from the metrological point of view.

While admitting that adoption of a single primary standard approach gives improved reproducibility (provided liquid junctions are always made in exactly the same way) Bates (1981) points out that this is at the expense of meaning, because the residual liquid-junction potentials become included in the values of the secondary standards. During the entire development of the NBS scale, Bates and his co-workers have maintained the primary objective of ensuring reproducibility of experimental pH measurements, but concomitantly, they have endeavoured to preserve as much meaning to the latter, in terms of conventional activity, as possible. A further powerful argument advanced by Bates in favour of the multiprimary standard approach centres on the expressed fear that endorsement of a new approach on an international scale could lead to a chaotic situation, for some nations have stated that they will not alter their present procedures.

The most recent stand taken by IUPAC (Covington et al. 1983) constitutes a compromise which recognizes the advantages of both the multistandard approach, and that based on the single standard with associated operational standards. The exact procedure and choice of standard reference solutions that are recommended, vary according to the degree of precision required in the measured pH and to the desire for interpretation in terms of hydrogen-ion concentration or activity.

Measurement of pH in practice

The appropriate place for measurements of pH(S) values of primary standards and pH(X) values of secondary standards is a national standards laboratory. This is where the pH scale is carefully maintained and preserved. Such safe custody is analogous to that of the platinum metre bar standard preserved in Paris. These standards are never removed from safekeeping in order to make practical measurements, whether these be of a routine nature, or in a carefully conducted research operation. Instead, the procedure is to make use of carefully chosen secondary standards. Thus, in the present section we are concerned, *inter alia*, with buffer solutions that are suitable for use as secondary pH reference standards.

In a limited number of practical situations, such as determination of equilibrium constants for reactions involving protons, the aim is to measure hydrogen-ion *concentrations*. Here, it is preferable to dispense with pH and to make direct emf/mV measurements following the approach described in Chapter 4. In this section, on the other hand, we are concerned with pH as a property of a solution which can be measured reproducibly with reference to an established scale, but which is limited to some extent in its interpretation. Most pH measurements, certainly in industry but also in other chemical, biochemical and in clinical applications, are concerned with the reproducibility (from place to place and from time to time) of the parameter designated as pH. Only a small fraction are made with a view to interpretation.

The required precision of pH measurements is tremendously variable covering the range from ±0.1 to ±0.001 on the pH scale. Nevertheless, in all electrometric measurements use is made of a cell of the type

$$\left. \begin{array}{c} H^+\text{-sensitive} \\ \text{electrode} \end{array} \right| \begin{array}{c} \text{solution} \\ \text{(either standard} \\ \text{(or unknown)} \end{array} \left\| \begin{array}{c} \text{reference} \\ \text{electrode} \end{array} \right. \quad (3.22)$$

The pH reading is given on a pH-meter which is really a potentiometer that senses the emf of the cell. The degree of sophistication of both the pH-meter and the vessel containing the standard reference or unknown solution is determined by the precision required in the measurement. pH-meters are discussed later in the present section. The solution-containing vessel could simply be an open beaker. Alternatively, when maximum precision is required, the vessel may be an elaborate thermostatted unit (see Figure 3.1) with a closely-fitting cap carrying the electrodes, thermometer and gas inlet/outlet. The last would be used for controlling the atmosphere in the vessel, for example with highly purified nitrogen in circumstances when it is necessary to preclude carbon dioxide and oxygen.

On rare occasions, a hydrogen electrode is used as the H^+-sensitive device. Whereas this is the most accurate electrode in this category, there are considerable disadvantages attached to its use in normal practice. For example, the platinum strip is readily poisoned by uncontrolled trace impurities; maintenance of a steady flow of hydrogen at a carefully regulated pressure requires more elaborate apparatus than is considered convenient in many applications; the pressure of readily reducible components in the solution could perturb the emf reading. Nowadays, a glass electrode is the most generally used hydrogen-ion sensor. In rare instances, quinhydrone and antimony electrodes have their place (Bates 1973) but the quality and reliability of modern commercially available glass electrodes make these last the most commonly preferred choice. Details of the chemical composition and design of both glass and reference electrodes needed to complete cells of the type (3.22) have been presented in Chapter 2.

In order to determine the pH of a given solution, the glass-electrode–reference-electrode–pH-meter assembly must be calibrated, using one or more reference buffer solutions of known pH. There is a need for regularly made calibrations because the response varies from one glass electrode to another and the response of a given glass electrode varies from one time to another. The reference solution must be a buffer, i.e. a solution which resists changes in pH upon addition of acid or alkali, so that its pH is not substantially affected by a reasonable amount of dilution or by contaminants which could inadvertently have been introduced from the original constituents or the environment. These buffers are classed as secondary standards; even if one happens to have an identical specification to a primary standard, it takes on the status of a secondary standard when used in a practical application as described in this section. When calibrating the apparatus, a buffer should be chosen which is as close in pH, composition, and concentration, to the unknown solution as possible.

The present section is divided into more detailed discussions of buffers (principles of buffer action and recipes for buffers needed in practice), pH-meters, measuring technique and measurements in solvents other than water.

Principles of buffer action

If about 1 cm^3 of either 100 mmol dm^{-3} strong acid or strong base is added to 1 dm^3 of pure water or water containing an inert electrolyte such as

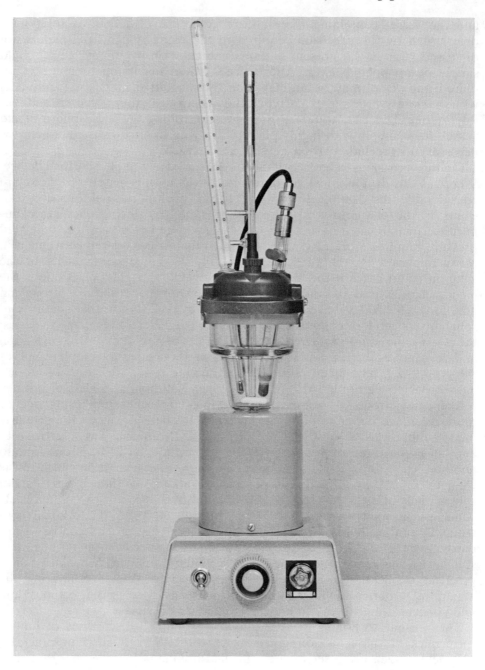

Figure 3.1 A gas-tight titration vessel; the inner glass vessel is normally surrounded by a glass jacket through which thermostatting water may be passed. The cap is fitted with a sealing ring and is clamped tightly on to the glass vessel. Five B14 tapered apertures in the cap are for carrying the electrodes, burrette tip, bubbler for inert gas and thermometer. Stirring of the solution is normally carried out magnetically (Metrohm AG, Switzerland)

Determination of pH values

sodium chloride, there results a marked change in pH to about 4 or 10, respectively. On the other hand, a similar addition made to ammonium acetate solution produces hardly any effect on the pH. Here, the ammonium acetate solution is behaving as a buffer; it shows the property of resisting a change in pH upon the addition of acid or alkali. Not only ammonium acetate, but other salts of weak acids and weak bases, act as buffers. Most buffers that are used in practice, however, consist of solutions each containing either a weak acid and its conjugate base or a weak base and its conjugate acid; counter-ions, of course, are also present.

The origin of buffer action is based on the equilibrium between the weak acid or base and the respective conjugate species. We explain this by reference to the weak-acid–conjugate-base type of buffer. The action of the weak-base–conjugate-acid type is governed by analogous principles. The equilibrium between a weak monoprotic acid, HA, and its conjugate base, A^-, may be represented simply by the reaction (3.23):

$$HA + H_2O = H_3O^+ + A^-. \qquad (3.23)$$

There might be additional side reactions to complicate the issue, but generally these also acquire equilibrium rapidly and are coupled to the equilibrium (3.23). Such side reactions do not affect the argument that follows. In a buffer solution based on the relevant weak-acid–conjugate-base system, the species HA and A^- are both present in significant concentrations at equilibrium. Upon addition of hydrogen ions to the buffer, these are removed through neutralization by A^-. Similarly, added hydroxyl ions are removed through neutralization by HA. Thus, the hydrogen-ion concentration, and hence the pH, is maintained more or less at a constant level. Chemical reversibility of reaction (3.23) and any associated side reactions is crucial to the ability of the solution to act as a buffer. The thermodynamic acid dissociation constant, $^T K_a$, of HA may be expressed as in equation (3.24):

$$^T K_a = \frac{m_{H^+} m_{A^-}}{m_{HA}} \cdot \frac{\gamma_{H^+} \gamma_{A^-}}{\gamma_{HA}} \qquad (3.24)$$

where each m denotes a molality value for the species indicated by the subscript and the γ's are the corresponding molal activity coefficients. The significance of m_{H^+} and γ_{H^+} are discussed earlier in this chapter. By combining equation (3.24) with the stoichiometric relations for the system, the condition of net electrical neutrality of the solution, and the Sørensen and Linderstrøm-Lang definition, (3.2), equation (3.25) may be readily derived:

$$pH = p\,^T K_a + \lg \frac{m_{A^-} + m_{H^+} - m_{OH^-}}{m_{HA} - m_{H^+} + m_{OH^-}} \cdot \frac{\gamma_{A^-}}{\gamma_{HA}}, \qquad (3.25)$$

where $p\,^T K_a = -\lg\,^T K_a$.

This equation may be used to calculate the pH of a buffer solution prepared from a given weak acid, HA, with known molalities of the weak acid and conjugate base. The activity coefficient, γ_{HA}, being that of an uncharged species, can be well approximated to unity. γ_{A^-} may be estimated using Kielland's (1937) approach. Nevertheless, the solution of equation (3.25) is not straightforward and for practical purposes some further useful approximations can be made. First, between pH values of 3 and 11, m_{H^+} and m_{OH^-} on the right-hand side become negligible. Secondly, if we regard the ionic strength as being fixed, the activity coefficients all assume constant values and we can invoke a conditional acid dissociation constant, K_a, defined by equation (3.26):

$$K_a = \frac{m_{H^+} m_{A^-}}{m_{HA}}. \tag{3.26}$$

Thus, for pH values restricted to the range between 3 and 11, and for a fixed ionic strength, equation (3.25) reduces to equation (3.27) which is much more manageable in calculating the pH of a buffer:

$$pH = pK_a + \lg \frac{m_{A^-}}{m_{HA}}. \tag{3.27}$$

In order to describe the effectiveness of a buffer in resisting pH changes, van Slyke (1922) introduced the idea of buffer capacity or buffer value, β. β is defined by equation (3.28):

$$\beta = d[B]/d\,pH, \tag{3.28}$$

where $d[B]$ is an increment of strong base added to the buffer, expressed as moles of hydroxide per dm^3. We see, then, that β represents the increment of strong base needed to change the pH of a buffer by a small amount and is the reciprocal of the slope of a pH-neutralization curve. An increment of strong acid, equivalent in effect to a negative increment, $-d[B]$ of strong base, produces a decrease in pH. Hence, $d[B]/d\,pH$ is always positive.

An expression for β in terms of the known properties of a buffer solution can be obtained by rearranging equation (3.27), making certain substitutions and differentiating. The result is equation (3.29) which, of course, is subject to the same restrictions as equation (3.27):

$$\beta = d[B]/d\,pH = 2.303 K_a c m_{H^+}/(K_a + m_{H^+})^2. \tag{3.29}$$

where c = the initial molality of weak acid
$\quad\quad\quad = m_{HA} + m_{A^-}$

An analysis of equation (3.29) leads to the following two conclusions:

(i) The maximum buffer capacity, β_{max}, is equal to 0.576C and is thus independent of K_a.
(ii) β_{max} corresponds to a buffer ratio, m_{HA}/m_{A^-}, of unity.

By applying conclusion (ii) to equation (3.27), we see that the solution acts most effectively as a buffer when the pH is equal to pK_a.

So far we have restricted our attention to buffers of constant ionic strength. We now consider the effect of adding a neutral electrolyte such as sodium chloride. The attendant increase in ionic strength changes the activity coefficients of electrically charged species, whereas the activity coefficients of neutral species are hardly affected at all. Thus, with our weak acid/conjugate base type of buffer, added salt would decrease γ_{H^+} and γ_{A^-} but γ_{HA} would be uninfluenced. From equation (3.25), therefore, it may be concluded that the pH would fall. Similarly, it may be shown that the pH of a weak-base–conjugate-acid type of buffer would rise upon adding a neutral electrolyte. Actual changes in pH vary with the K_a value, the concentration of the weak acid or base and the buffer ratio (e.g. m_{A^-}/m_{HA}) as well as with the concentration of added neutral electrolyte. Observed or calculated changes on the pH scale cover the range (approximately) ± 0.007 to ± 0.1 for neutral salt concentrations between 10 and 100 mmol kg^{-1} (Bates 1973).

Another factor that must be considered is dilution of a buffer. The dilution value, $\Delta pH_{1/2}$, is defined as the increase in pH suffered by a solution of initial concentration, c_i, on dilution with an equal volume of pure water. The dilution value can be determined experimentally or calculated from theoretical considerations. For the latter, the reader is referred to the book by Bates (1973). For a weak-acid–conjugate-base type of buffer, $\Delta pH_{1/2}$ typically ranges from about 0.009 to 0.024 for initial concentrations between 5 and 100 mmol dm^{-3} (Perrin and Dempsey 1974).

Finally, in this section, we consider substances known as self-buffers. Certain single compounds when dissolved in water to a specific concentration establish a reproducible pH. Ammonium acetate, as referred to at the beginning of the section, is an example of such a self-buffer, although it is not such a useful one because the buffer capacity is small. Well-known useful examples of self-buffers are potassium hydrogen phthalate and sodium tetraborate (borax). As discussed in this chapter, potassium hydrogen phthalate is taken as the single primary standard for defining the British pH scale. Both potassium hydrogen phthalate and borax are used as primary standards in the US multiprimary standard system. The self-buffering action of potassium hydrogen phthalate is based on the fact that phthalic acid is dibasic with pK_a values at 25 °C of 2.95 and 5.41. Thus, potassium hydrogen phthalate is a partially neutralized salt which in aqueous solution dissolves as a strong electrolyte into potassium cations and hydrogen phthalate anions. The latter species is both the conjugate base of phthalic acid and the conjugate acid of the phthalate dianion. Thus, a certain amount of hydrolysis takes place in solution resulting in a reproducible pH approximately midway between the two pK_a values of phthalic acid. Borax dissociates in solution to form principally metaboric acid, HBO_2, and metaborate ion, BO_2^- in almost equimolar proportions. These two species constitute a weak acid/conjugate base pair and hence form a buffer system. There are slight complications arising from the tendency of boric acid and borate ion to form polymers.

Recipes for standard buffers

Buffers can be purchased, and in practice usually are. The interested reader can, however, find recipes in the literature (Bates 1973; McGlashan 1971; IUPAC 1979).

Procedures for routine pH measurements

In this section, we are concerned with the use of glass-electrode–reference-electrode–pH-meter assemblies. The resistance of a glass electrode is extremely high, being of the order of 100 MΩ. Therefore, to measure the emf of a cell such as (3.22), an ordinary voltmeter cannot be used if one is to avoid perturbing the measurement. Instead, use must be made of a potentiometer with a very high input impedance (i.e. a low bias current). Modern pH-meters fulfil this requirement in that typical input impedances are around 10^{13} Ω. A circuit in the instrument amplifies the emf and electronic adjustments of one kind or another are made to the signal prior to display of the result. These instrumental adjustments provide the user with a choice between displaying the result as either emf/mV or pH, allow setting of the appropriate Nernstian slope factor, $RTF^{-1} \ln(10)$ (refer to equation (3.15)), and also permit corrections to be applied for imperfections in the electrodes. Nowadays, both analogue (meter) and digital displays are available. Digital displays have the advantages of being simple, exact and unequivocally readable, the need for interpolation between graduation marks being automatically eliminated. Of course, even though several digits (e.g. four) may be displayed, not all are necessarily significant. Digital displays are becoming increasingly popular, although certain advantages of analogue readouts remain. For example, the rate of change of pH, as in a titration, is more satisfactorily indicated by travel of a needle than by rapidly changing digits.

If glass electrodes were to behave perfectly, one ought, ideally, to be able to obtain a pH reading of a test solution as follows

   (i)   Select a suitable standard (secondary) buffer with a pH as close to the test solution as possible.
  (ii)   Check that the test solution and buffer are at the same temperature; set the temperature compensation control of the pH meter to this value. (This sets the Nernst slope factor, $RTF^{-1} \ln(10)$, to the correct value.)
 (iii)   Place a portion of the buffer in the vessel of the glass-electrode–reference-electrode pH meter assembly and adjust the standardization control so that the display reads the nominal pH of the buffer.
 (iv)   Replace the buffer by the test solution, rinsing the electrodes appropriately beforehand. Read the pH.

Should the measurement of the pH value be required to within only ±0.05, the above four steps are generally sufficient. When better precision is required, however, a more elaborate procedure is necessary, involving adjustment of other pH-meter controls in order to compensate for electrode imperfections and perhaps utilizing more than one reference buffer in the standardization step. When aiming for the higher precision measurements, account must be taken of the facts that the response varies from one glass electrode to another and also that for a given glass electrode the response varies with time. Moreover, many glass electrodes exhibit sub-Nernstian behaviour; that is, the observed slope of the E versus pH line is smaller than the theoretical slope, $RTF^{-1}\ln(10)$. The aim of manufacturers is to produce glass electrodes which have exactly the same potential as a saturated calomel reference electrode (SCE) when immersed in a solution of defined pH 7.000 and at all temperatures.

The observed pH at which the emf between a given glass electrode and an SCE is 0 mV is termed the *electrode isopotential point.* Thus, ideally, the electrode isopotential point would be equal to 7.000 and independent of the temperature. In practice, the isopotential point is rarely at exactly 7.000 and usually shows a slight temperature variation. Compensation for both sub-Nernstian response and for deviation of the isopotential point from the ideal value may be effected by a series of alternate adjustments of the slope control and the zero (isopotential) control. With a little experience, an operator can aquire adroitness in using these controls to provide a pH readout that closely follows the actual electrode response. The slope control produces the same effect as the temperature compensation control in that the value of $RTF^{-1}\ln(10)$ is electronically adjusted. For high-precision work, both the temperature compensation and slope controls are needed and can be regarded as being complementary to each other. The zero control is really a zero offset which enables the isopotential point of the meter (i.e. the pH reading corresponding to an emf of 0 mV) to be set at any desired value. For the most effective use of these controls the user is referred to the manufacturer's instructions. A detailed description may also be found in the book by Westcott (1978).

Once the zero and isopotential controls have been set correctly, the glass-electrode–reference-electrode–pH-meter assembly may be calibrated by using either a single standard buffer or two such buffers. The choice between these two approaches is dictated by the conditions of the measurement. If pH measurements are to be made on a series of samples, such as blood, covering a narrow pH range, the single-buffer calibration approach is adequate. On the other hand, if the samples cover a wider range of pH, the two-buffer approach is preferred. In the single-buffer method, the procedure is carried out along the lines (i) to (iv), given for the ideal situation described above. Here, the E versus pH line is assumed to have the Nernstian slope or the value arrived at by adjustment of the zero and slope controls to compensate for non-ideal behaviour of the electrodes. The buffer should be carefully chosen so that its pH approximates as closely as possible to the test solutions. If appropriate, the reliability of the measurement can be greatly enhanced by choosing the composition and ionic strength of the buffer to resemble those of the test solution. With the two-buffer method, one of these is used for standardization and the other for span adjustment through further adjustment of the slope control. The two buffers must be chosen so that their pH values bracket those of the test solutions.

Nowadays, pH-meters are available with an automatic temperature-compensation facility. A thermistor or resistance thermometer probe is immersed in the test solution and any temperature fluctuation results in automatic adjustment of the emf versus pH line slope. This facility is particularly useful when monitoring an industrial process stream. It must be appreciated, however, that the attainment of equilibrium is not necessarily instantaneous.

One of the latest developments on the market comprises pH-meters with microcomputer control devices. With these instruments, standardization using either the one- or the two-buffer method is automatic. Furthermore, automatic adjustment is effected of the temperature, slope and zero (isopotential) compensation devices. Data and parameter values can be printed at precisely predetermined times. Most attractive features are the light weight and compact construction of these new pH-meters.

pH measurements in solvents other than water

Here, we concern ourselves briefly with situations in which the solvent may be placed in any of the following categories:

(i) Partially aqueous solvents, that is, mixtures of water and some organic solvent that is miscible with water, such as an alcohol, dioxane or acetone.
(ii) Non-aqueous solvents such as acetic acid, acetone, acetonitrile, dimethylformamide, pyridine.
(iii) Deuterium oxide (D_2O, heavy water).

The acid–base behaviour and strength as well as the activity coefficient of a given solute depend upon the acid–base and other properties, such as relative permittivity, of the solvent. Consequently, the pH scale established for water is not quantitatively applicable to other solvents and direct relationships between pH values for a given solute in different solvents are not readily derivable (Marcus 1980). Accordingly, the general approach is to establish a pH scale for every particular solvent using assigned pH(S) values in an analogous way to the system established for water (Bates 1981). The work involved in setting up buffer standards for each and every solvent system is a formidable task and at present is far from being complete. Most of the effort to date has involved methanol, ethanol and their aqueous mixtures. Tables of pH standards for a limited number of solvent systems may be found in the book by Perrin and Dempsey (1974).

Glass electrodes respond in a reproducible way to protons in the solvents categorized in this section, provided care is taken to keep the glass membrane hydrated. This generally involves occasional soaking in water. One drawback with solvents in categories (i) and (ii) (above) is that glass electrodes, in some cases, respond slowly.

As far as the reference electrode of the measuring assembly is concerned, solvents in categories (i) and (ii) tend to introduce large liquid-junction potentials. Westcott (1978) describes ways of reducing these. Perrin and Dempsey (1974) point out that the liquid-junction potentials can, with proper care, be made reproducible, in which case they present no obstacle to the reproducible determination of pH in the solvents concerned as long as both standard and unknown have the same solvent composition.

With respect to heavy water, glass electrodes respond satisfactorily to deuterium ions. A pD scale has been established using assigned values for three different standards (Paabo and Bates 1969). If it so happens that a measuring assembly is standardized against an aqueous pH buffer, the relationship in equation (3.30) can be used to obtain pD (Glascoe and Long 1960; Covington et al. 1968):

$$pD = \text{meter reading} + 0.40 \qquad (3.30)$$

The interpretation of pH

The principles described in earlier sections provide a common basis for all pH measurements. Using these principles, pH may be measured precisely and reproducibly from time to time and from place to place. Comparisons between pH values obtained in different laboratories are accordingly significant.

The interpretation of a measured pH value, however, is subject to some uncertainty. The uncertainty originates primarily from our inability to determine single-ion activities. While the operational definition of pH is aimed at ensuring dominance of the reproducibility aspect, the Bates–Guggenheim convention, in fact, provides as much quantitative meaning in terms of chemical equilibria as possible. Indeed, Bates and Guggenheim (1960) point out that the relationship, (3.31), holds with an accuracy of ±0.02 on the pH scale for all aqueous solutions of ionic strength not exceeding 100 mmol kg^{-1} and in the range

$$2 \leqslant pH \leqslant 12$$
$$pH = pm_{H^+}\gamma_{H^+} \qquad (3.31)$$

where $pm_{H^+}\gamma_{H^+}$ is determined conventionally in the manner described, using equation (3.20). Thus, within the limits of applicability described above, pH may be interpreted with the degree of uncertainty deduced by Bates and Guggenheim. In this sense, therefore, the Bates–Guggenheim convention is most judiciously chosen.

Systems of high ionic strength

Two examples of systems in this category are body fluids (such as blood) and sea water, with ionic strengths of 160 and 700 mmol kg^{-1}, respectively. Both of these ionic strength values are beyond the range of applicability of the Bates–Guggenheim convention, rendering the interpretation of pH measurements subject to considerably greater uncertainty than discussed above.

In the case of body fluids, the pH averages around 7.4 and the variation is rather small. What is required is a discrimination between small pH changes in order to seek correlations with pathological conditions and physiological processes. The interpretation of pH in terms of proton activities or concentrations is of only minor interest in this context. In order to obtain precise and reproducible measurements of pH, two standard buffers based on potassium dihydrogen phosphate and disodium hydrogen phosphate in the proportions 1:3:5 and 1:4, respectively, have been devised (Bates 1973). These are assigned defined pH values at 37 °C. The ionic strength of the buffers is relatively high in order to minimize residual liquid-junction potentials.

It has been demonstrated that sea water behaves as a 'constant ionic medium', stabilizing the activity coefficients of solutes present in small concentrations and reducing or eliminating the liquid-junction potential encountered in measuring the pH of sea-water systems (Hansson et al. 1975; Khoo et al. 1977). These favourable characteristics justify the formulation of a sea-water pH scale based on hydrogen-ion molality (instead of activity) and defined by the relationship

$$pm_H = -\lg (m_H)_X \qquad (3.32)$$

where m_H = the molality of 'free' hydrogen ion in the sea water, X (Hansson et al. 1975; Khoo et al. 1977; Bates 1982). Buffers made up in synthetic sea water which have been suggested as possible pm_H standards for marine systems are equimolal Tris/Tris · H^+, Bis/Bis · H^+, 2-AMP/2-AMP · H^+ and Morph/Morph · H^+, where

> Tris = tris(hydroxymethyl)aminomethane,
> Bis = 2-amino-2-methyl-1,3-propanediol,
> 2-AMP = 2-aminopyridine,
> Morph = morpholine,

(Ramette et al. 1977; Bates and Calais 1981; Czerminski et al. 1982).

Concluding remarks

Bates has recently written "One may well despair of ever finding a concise and fully adequate answer to the question 'how is pH defined?' " (Bates 1981). Nevertheless, despair and inadequacy ought not to lead to disillusionment and discouragement. We may liken the situation to that of our understanding of the force of gravity. It varies from country to country but all scientists agree that the more one understands the theories underlying the phenomenon and standardizes one's approaches, the more precisely the absolute value becomes established.

The pH scale has progressed a long way since its inception in 1909 and

Determination of pH values

sophisticated improvements are still being made as well as some fine tuning to the theories. Industry and medicine can feel most grateful to those who have kept the subject alive and modern.

A different range of benefits accrue from using the concentration, as distinct to activity, scale—a branch of potentiometry which has evolved far more recently than the pH scale—and this is detailed in the next chapter.

4

Determination of hydrogen-ion concentrations

> 'Chemists underestimate the amount of information obtainable from emf titrations.'
>
> L. G. Sillén

Data

Good data will often outlive the researchers who produced them. Many physical chemists still undertake recalculations based upon good data which are many decades old. The building blocks of an earlier hypothesis are dismantled and reconstituted into a more modern theory.

As suggested by the late Professor Sillén's quotation above, titrations can contain much more information than has been traditionally utilized (Brauner et al. 1969). Raw potentiometric data have been treated and condensed into so-called 'constants'. It is particularly important that such constants are unambiguous and internationally compatible. Unfortunately, this has not always been the case; for example, many published constants mix hydrogen-ion *activities* with species *concentrations*. Such data are not as useful as they might be as input to computer speciation models and it is preferable to recompute these constants.

Similarly, good data are dependent upon good electrode calibrations regardless of whether one is measuring activities, a_{H^+}, or concentrations, $[H^+]$, of hydrogen ions. Such calibrations, as with all data acquisition, ought to be performed as cost-effectively as possible; in practice this means that some parameters are measured each time a titration is performed, some are measured less frequently, e.g. weekly, and some are acquired from the literature or from calibrations performed many months earlier. The philosophy and principles used to apportion these parameters among these three categories in order to optimize both precision and cost-effectiveness is both an art and a science. This chapter considers these features in respect of hydrogen-ion-concentration measurements.

The chapter particularly focuses upon the measurement of hydrogen-ion concentrations rather than activities (as studied in Chapter 3) because the former are needed for speciation studies using models as described in Chapter 5.

Determination of hydrogen-ion concentrations

The fundamental objective is the calibration of a glass/reference electrode cell

| first reference half-cell (in glass electrode) | glass membrane | test solution | liquid junction | second reference half-cell (external) |

over the range of [H$^+$] likely to be encountered in the experiment proposed (for example, the pK determination using titrations). As the [H$^+$] is varied the activity coefficient, y$_{H^+}$, will undoubtedly vary unless the ionic strength (I) of the solution is maintained constant using an ionic background electrolyte.

Equation (3.8) has shown a relationship between activity coefficient and ionic strength, I. Over the last four decades many formation constants have been obtained at a wide range of different temperatures and ionic strengths. For example, the Sillén school in Stockholm has reported numerous results obtained at 25 °C and 3 mol dm^{-3} sodium perchlorate, under which conditions Biedermann (Williams 1976) has demonstrated that all activity coefficients are held constant, provided the sum of the equivalent concentrations of all the positive and negative ions disappearing during a complex forming reaction does not exceed 150 mmol dm^{-3}. There is, however, another trend which is prevalent, namely to use 150 mmol dm^{-3} sodium chloride or perchlorate at 37 °C, corresponding to blood plasma conditions. The latter are pertinent to many of the examples of speciation modelling discussed in Chapter 5. One disadvantage of choosing 150 mmol dm^{-3} as the concentration for the background electrolyte is that the change in ion concentration accompanying a complex forming reaction is limited to a maximum of 8 mmol dm^{-3} without significantly affecting activity coefficients (Williams 1974).

Under ideal conditions when such ionic backgrounds are successful at holding activity coefficients constant, a Nernstian relationship holds true (Brauner et al. 1969):

$$E = E_{\text{const}} + s \lg [H^+] + j_H[H^+] + j_{OH}K_W/[H^+]. \qquad (4.1)$$

In equation (4.1), $E_{\text{const}} = E_{r_2} - E_{r_1} - E_g^0 + s \lg y_{H^+}$, where E_{r_1} and E_{r_2} are the electrode potentials of the first and second reference half-cells, respectively; E_g^0 is the standard potential of the glass membrane at unit activity of hydrogen ions; y$_{H^+}$ is the molar activity coefficient of hydrogen ions, and s is the Nernstian slope of the cell which should have a value approximating very closely to $2.303RTF^{-1}$ (R being the gas constant, F the Faraday constant, and T the absolute temperature).

Clearly, a good calibration procedure performed over a range of [H$^+$] ought to produce constant values for E_{const} and s, the latter being expected to have a value of 59.157 mV per lg [H$^+$] at 25 °C. In practice, the j_H and j_{OH} terms are usually negligible within the approximate $-\lg[H^+]$ range 2–10.

In reality, the E_{const} values have to accommodate a wide range of problems such as I varying slightly during a titration as anions and cations combine and reduce the m_i term in equation (3.7), and the fluctuating manner in which glass electrode properties vary from day to day. The 'art' of a good calibration procedure is to arrest, or to quantify, each of these idiosyncrasies as well as possible such that emf values from the working solution can be converted into [H$^+$] to within a precision of ± 0.02 mV. Clearly, since some of the above electrode characteristics have temporal or environmental properties, calibration of the electrodes *within the actual working titration* is a highly desirable aim.

Strong-acid–strong-base procedures

If one assumes that one has access to solutions of a strong acid such as hydrochloric or perchloric acid, and of a strong base such as sodium or potassium hydroxide (in reality it is not easy to obtain analytically precise figures for such solutions—many cross-titrations which exclude carbon dioxide are necessary) one may set up a series of cells and correlate E with [H$^+$]. The simpler and more cost-effective way of achieving this is, of course, to arrange a stepwise titration of acid into base or *vice versa*. Each step in this titration is, in reality, a new cell of a different [H$^+$] to the previous one. Providing the introduction of the burette and any tip leaking do not introduce strong titrant, and providing the stirring is efficient, one may thus obtain a whole range of calibration points in a matter of an hour or two. Once again, carbon dioxide must be scrupulously excluded.

As with all acid–base titrations, there is a gradual change in emf with amount of titrant added and then a sudden change as the 'end-point' is reached, followed by a gradual change plateau reached again after the endpoint. Calibration titrations may be contrasted with acid–base determination titrations in that the former require the points on the plateaux either side of the endpoint whereas the latter are primarily interested in the points surrounding the endpoint.

Strong-acid–strong-base titrations have a limited usefulness for many reasons:

 (i) Unless there is sufficient excess acid or alkali present to ensure that the solution is concentration-buffered, small errors become very significant, for example, the loss of some H$^+$ which becomes absorbed onto the glass of the electrodes.
 (ii) Glass electrodes do not behave well in the alkaline regions because they are somewhat sensitive to alkaline metal ions; this is especially so above $-\lg[\text{H}^+] = 11.0$.
 (iii) At the acid end of the pH scale the effects of acid upon liquid-junction potentials is also important.

Reasons (ii) and (iii) are taken into account by the j_{OH} and j_H terms in equation (4.1). Thus, strong-acid–strong-base electrode calibrations are only really usable in the $-\lg[\text{H}^+]$ ranges of about 2.3–2.9 and 10.8–11.3. Sadly, most working titrations where one is researching metal–ligand complexing, for example, are within our calibration 'blind spot' of $-\lg[H^+]$ 2.9–10.8 and so supplementary titration approaches are necessary.

Internal calibration procedures

Techniques commonly used include replacing the concentration-buffered range with more conventional buffering by titrating a weak acid (or base) with a strong base (or acid) and taking calibration readings in the buffered region of the titration, due allowance being made for the pK of the weak acid (or base) present. Indeed, this concept may be extended further in that the aforesaid pK itself may be simultaneously determined within this titration. This is termed 'internal calibration' as the E_{const} and s parameters are measured within the test solution itself. Such an approach overcomes problems arising from the slight irreproducible aspects of liquid-junction potentials from calibration solution to working solution and variations of potentials with time (May et al. 1982).

At first sight it may seem surprising that calibration and equilibrium constants can be determined from the same set of data, the weak acid (or base) acting as both 'referee' and 'contestant' but, providing a reasonably large range and number of titration points can be acquired, we can reason as follows.

Let us consider the stepwise titration of a ligand whose anion, L^-, is monoprotic. Suppose the ligand solution is acidified with a mineral acid and that a strong base, SB, is titrated into the solution.

Let $(T_L)_{init}$ be the initial total concentration of the ligand in the titration vessel; $[MA]_{init}$ be the initial concentration of mineral acid in the titration vessel; $[SB]$ be the concentration of strong base titrant; $[LH]$ be the current concentration of protonated ligand for a given point in the titration; $[L^-]$ be the current concentration of ligand anion for the same given point in the titration; $[H^+]$ be the current concentration of free, hydrated, hydrogen ions, i.e. those not bound to L; V_{init} be the initial volume of the acidified solution in the titration vessel; V_t be the titre volume of SB added at any stage in the titration; K_W be the value of the ionic product of water, corresponding to the background medium and the temperature of the titration; β_{101} be the protonation constant of the ligand. (β_{pqr}, in general, is the nomenclature used to denote the overall formation constant for the complex, $L_pM_qH_r$, r being negative for OH complexes; clearly; $\lg \beta_{101}$ is equal to pK for the ligand.)

Suppose the titrand solution is made up from the ligand in its protonated form, LH, then for a given point in the titration, the following equations may be used to describe the proton–ligand equilibrium and the electrode response:

$$E = E_{const} + s \lg[H^+]. \quad (4.2)$$

(As given earlier as equation (4.1) but avoiding the more concentrated acid or alkaline regions where j_H and j_{OH} need to be introduced.)

$$\beta_{101} = [LH]/[L^-][H^+] \tag{4.3}$$

(by definition),

$$\frac{(T_L)_{init} \times V_{init}}{V_{init} + V_t} = [L^-] + [LH] \tag{4.4}$$

(mass balance in ligand),

$$\frac{(T_L)_{init} \times V_{init} + [MA]_{init} V_{init} - [SB] V_t}{V_{init} + V_t} = [H^+] + [LH] \tag{4.5}$$

(mass balance in protons).
If there are n points in the titration, we have $4n$ equations of the type (4.2)–(4.5).

Amongst these $4n$ equations, $(T_L)_{init}$, $[MA]_{init}$ and $[SB]$ are known from the analytical concentrations of the experimental solutions. V_{init} is known from the details of making up the titrand solution. V_t and E are measured at each point. K_W is known from the literature or from earlier measurements in one's own laboratory. There are three unknowns which apply throughout the titration, namely, E_{const}, s and β_{101}. In addition, the three unknown variables, $[H^+]$, $[L^-]$ and $[LH]$ have particular values at each point in the titration.

It follows that amongst the $4n$ equations, there are $3 + 3n$ unknowns. In principle, therefore, providing $n \geq 3$, the equations can be solved simultaneously to yield values for the unknowns.

In practice, several dozen titration points are usually taken and so least-squares 'best' solutions for E_{const}, s, β_{101}, etc. are easily obtained.

Assessment of analytical errors

It is generally true that all experimental data include both random and systematic errors. The former are minor experimental or accidental errors arising from outside disturbances (for example, a temperature change), mildly fluctuating conditions, or small errors of judgement near the limits of detectability or readability of the measuring device. The frequency of these random errors is symmetrical about the mean result of a large system of measurements. Systematic errors, on the other hand, are constant errors in which all the results are incorrect in the same direction and by approximately the same amount. Such factors as faulty calibration, personal prejudices in how the data are collected, or a systematically incorrect technique can all produce systematic errors (Pantony 1961).

In potentiometric work there is always some overlap between these two types of error but, nevertheless, they do not necessarily need to appear in the final constants since random errors can be quantified and allowed for providing a sufficiently large number of data points are recorded and many systematic errors can be allowed for, providing that they are acceptably low in comparison with the data measured, by building such errors into corrected analytical concentrations, $[A]_{init}$, $[MA]_{init}$, etc., volumes, V_{init}, or constants such as K_W. This is equivalent to considering these systematic errors as additional unknowns and treating them as part of the overall objective of producing a better fit of data to mass balance equations by obtaining E_{const}, s and β values.

Naturally, one must retain at least one of the quantities as a fixed reference point, otherwise the arithmetical solutions of the simultaneous equations become wild and may produce spurious fits of values to the experimental data. In practice, the situation may be influenced by the mathematical approach or method used to find the solutions; some demand that a certain minimum number of experimental quantities are used as fixed reference points.

Mathematical methods used for solving non-linear simultaneous equations employ a range of iterative numerical procedures, which are code-named Newton–Raphson, pit-mapping, Gauss–Newton and Nelder–Mead (or simplex extended) methods. Gans (1976) has critically reviewed and assessed these approaches and drawn attention to their use *inter alia* in the chemical problems involved with complexing equilibria.

Programs available

Just as glass electrodes are usually purchased commercially rather than 'home-made', so too most glass electrode users take already available programs rather than compose their own suite of computer software for solving these simultaneous equations. One of the aims of this chapter is to inform the reader about the range of, and some pitfalls in, programs widely available.

The original program in this area was LETAGROP, developed in Sillén's school in Stockholm (Sillén 1962). It is a very large and versatile program which can embrace many problems in chemistry (ranging from gases through solutions to solids) and in other physical sciences. Since 1962, sophisticated large-memory desk calculators have been developed and so now a wide range of programs is available, from advanced main-frame computer techniques to simpler hand-calculator methods (ranging from the luxurious to utilitarian approaches).

Some of these programs are now described, before our own experiences are recounted.

1. MAGEC

General

MAGEC (Multiple Analysis of titration data for Glass Electrode Calibration) optimizes, simultaneously, any or all titration parameters pertinent to the calibration of glass electrodes either during a strong-acid–strong-base titration or during a ligand-protonation titration (May et al. 1982).

The protonation constants of a ligand and the glass-electrode parameters can be determined simultaneously along the lines indicated on the previous pages. The program contains a subroutine CALIBT which permits strong-acid–strong-base data to be used to refine the value of pK_W and to monitor electrode performance and quality from titration to titration.

In addition, when processing ligand-protonation titration data it permits the calculation of E_{const} for the exact pH range of the titration and especially the highly buffered regions. Ligand, acid, and base concentrations can be refined as well as electrode slope, s, and the initial volume of titrand.

Principles of MAGEC

The MAGEC program as listed in the Appendix is based upon the relationship of equation (4.2). In circumstances where strong-acid–strong-base titrations are to be used in the calibration of an electrode system, MAGEC uses a subroutine entitled CALIBT. CALIBT first analyses the data by the method of Gran (Gran 1950, 1952; Rossotti and Rossotti 1965). Since the potentiometric data are transformed into a linear form, this analysis gives a good indication of glass-electrode performance and also yields an end-point that is independent of the intercept used in equation (4.2).

Further processing of strong-acid–strong-base titrations is divided into three stages. To begin with, the input concentration values are used to calculate free hydrogen-ion concentrations at each point, and a linear least-squares fit is performed on the data before and after the endpoint and over the entire range. Because of the effect of relatively small errors in the concentrations of titrant and titrand, the least-squares straight line does not normally possess a Nernstian slope, so the concentration of the titrand is varied slightly until the slope from the data before the end point becomes Nernstian. It is then possible to adjust the titrant concentration in a similar manner, but on the basis of the whole range of data.

A factor that critically affects the refinement of the titrant concentrations in the final stage is the value used for the dissociation constant of water, K_W. Either the titrant concentration or K_W (but not both) can be determined by finding the value which yields the most ideal least-squares slope. To accommodate those situations in which K_W is uncertain, CALIBT permits the user to vary the estimate systematically. The recommended procedure for using CALIBT is illustrated in Figure 4.1.

Data obtained in a ligand protonation titration are processed by the main routine of MAGEC. Providing a sufficiently large number of titration points have been taken, equations (4.2) to (4.5) may be solved to yield, simultaneously, E_{const} and the ligand protonation constants. The MAGEC analysis involves general optimization of parameter values by using an objective function based on titration volumes. The sum of squared residuals is minimized by the simplex method described by Nelder and Mead (1965). In fact, any of the parameters, E_{const}, s, the β-values, T_L, T_H and K_W can be

Determination of hydrogen-ion concentrations

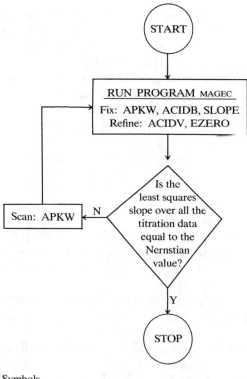

Symbols

APKW = apparent dissociation constant of water (K_w)
ACIDV = titrand acid concentration (negative for alkali)
ACIDB = titrant acid concentration (negative for alkali)
EZERO = electrode intercept (E_{const})
SLOPE = electrode slope (s)

Figure 4.1 Flow chart for MAGEC program

flagged for refinement, so the specific procedure is left largely in the hands of the user.

MAGEC in practice

In our experience, the calibration of an electrode system is best carried out in the following steps.

(i) Even when the calibration is needed outside the pH ranges 2.3–2.9 or 10.8–11.3, it is recommended that an initial estimate of E_{const} be obtained by applying a CALIBT analysis to the titration data for a strong acid versus a strong base. An additional advantage of this step stems from the possibility of using the CALIBT analysis for checking the reliability of an individual glass electrode. Lifetimes of glass electrodes differ considerably and it is often difficult to detect the first signs of deteriorating performance. In the MAGEC analysis of strong-acid–strong-base titration data, however, the appearance of sudden and marked increases in the titration-volume residuals signals the imminent demise of the electrode. A second useful criterion of incipient electrode unreliability is a tendency for discrepancies to develop between the endpoints determined by CALIBT minimization of the Nernstian response and by Gran plot.

(ii) The E_{const} value obtained in step (i) should be used in conjunction with data obtained for titration of a ligand (perhaps acidified with mineral acid) with a strong base, in a suitable program for refining the values of the protonation constants for the ligand. The authors have used MINIQUAD (Sabatini et al. 1974; Gans et al. 1976) for this purpose for several years. The recent report of SUPERQUAD (Gans et al. 1983; Vacca and Sabatini 1984) places model selection on a sounder statistical basis. Individual titrations (i.e. either replicate titrations or titrations differing in the total concentration of ligand) should be processed together to yield global values for each protonation constant.

(iii) The ligand-titration data and the protonation constants from step (ii) are processed by the main routine of MAGEC to yield refined E_{const} values.

(iv) Steps (ii) and (iii) are repeated cyclically until convergence to a satisfactory degree is obtained.

(v) If desired, the values of E_{const} and the protonation constants may be improved further by refining the values of T_H, T_L and K_W in a series of subsequent cycles. Care must be taken, however, to ensure that the calculation is not performed with more degrees of freedom than are warranted by the accuracy of the data. If this precaution is neglected, all optimization procedures can give rise to spurious results.

The number of parameters which can be refined with safety from data with a given experimental precision depends on the particular system being investigated. Thus, it is important to ensure that the refinement includes only those parameters for which optimization leads to a sufficient improvement in the sum-of-squares function that is to be minimized.

The flow diagram of Figure 4.2 illustrates steps (ii)–(iv).

Worked examples using MAGEC and MAGEC–MINIQUAD cycling

Example 1

An example of the calibration of a glass electrode/silver–silver chloride electrode pair by a strong-acid–strong-base titration.

First, for a given glass electrode and silver–silver chloride electrode pair a buffer line using five buffers established that the response was linear throughout the working pH range (see Table 4.1) at 25.0 °C.

Then a mixture of hydrochloric acid and sodium chloride was titrated with

Determination of hydrogen-ion concentrations

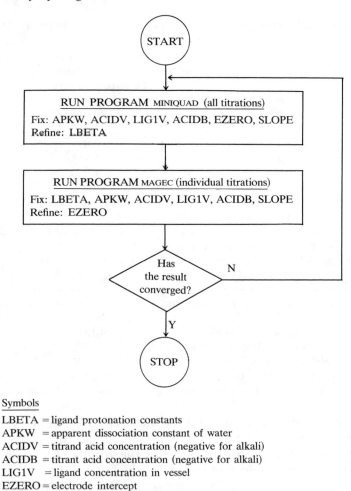

Figure 4.2 Flow chart for MINIQUAD–MAGEC cycling approach

a mixture of sodium hydroxide and sodium chloride. The exact concentrations were as listed in Table 4.1, sufficient ionic background sodium chloride being used to make $I = 1.000$ mol dm^{-3} in Cl$^-$.

The results of the Gran-plot analysis performed by the MAGEC program are also presented in Table 4.1. Considering the straightest segments of the plots, the endpoints obtained for the acid and alkaline pH data were 9.13(42) and 9.12(25) cm^3 respectively. The concordance between the latter two results indicates the absence of detectable carbon dioxide contamination in the sodium hydroxide titrant.

Table 4.2 reports the CALIBT procedure for optimizing acid and alkali concentrations. It may be seen that of the 59 points used in the present example, less than one third (i.e. 18 points) gave residuals greater than

Table 4.1 Strong-acid–strong-base titration (temperature 25 °C)

BUFFER LINE	Buffer pH	Electrode response (mV)	Deviation from best straight line drawn by inspection (mV)	Slope of buffer line
	1.68	−335.8	0.6	
	4.01	−193.7	−2.1	
	6.84	−27.2	0.8	59.89 mV/pH
	7.00	−18.7	1.9	
	9.18	+115.6	−1.9	

INPUT VALUES		
Concentrations (mmol dm^{-3}):	HCl, 19.2; NaOH, 42.12	
Initial volume of HCl:	20.00 cm^3	
pK_W:	scan between 13.89 and 13.92	
Electrode intercept:	400 mV	
Electrode slope:	59.16 mV per lg [H$^+$]	
(Expected endpoint:	9.12 cm^3 NaOH)	

GRAN PLOTS	Endpoints (cm^3 NaOH)	For points before the endpoint	For points after the endpoint
The straightest segment gives		9.1342	9.1225
The minimum standard deviation gives		9.1301	9.1253
The weighted average gives		9.1346	9.1726
The actual average gives		9.1356	9.1897

0.1 mV. This is an excellent check on the quality of the electrode system. Finally, by appraising several scans one obtains an estimate of E_{const} and a 'best' value for pK_W—Table 4.2 precises the results of six scans; the scans numbered 2 to 6 have constant values for the overall standard deviation and the electrode slope reaches the expected Nernstian value after sodium hydroxide concentration adjustment in all six scans. Thus, since we have an equal choice for 'best' results from scans 2 to 6, but note that the nominal acid and base concentrations are least adjusted in scan 3, then the 'best' values for pK_W and E_{const} are chosen as 13.90(2) and 480.2 mV respectively.

Example 2
The internal calibration of a glass electrode/silver–silver chloride electrode pair using a titration of a ligand solution and MAGEC-MINIQUAD cycling.
   The same electrodes were used as in example 1. Oxalic acid was titrated with sodium hydroxide, $I = 1.000$ mol dm^{-3} Cl$^-$ and at 25 °C. The input (IN) and output (OUT) data are summarized in Table 4.3. The initial value used for E_{const} was that obtained in example 1. Initial estimates of the lg β values were taken from the literature.
   In the first run, MAGEC refined E_{const} yielding a value which was then used in MINIQUAD. The latter yielded lg β values which, when used subsequently in MAGEC, gave a considerably lower objective function (showing an improved fit of the parameters to the data), although E_{const} happened not to have been

Determination of hydrogen-ion concentrations

Table 4.2 pK_W scans and fitting of titration points to Nernstian equation

Scan no.	pK_W value	Stage of CALIBT optimization*	Deviations of optimized from nominal concentrations (%) HCl	Deviations of optimized from nominal concentrations (%) NaOH	Least-squares values using all the buffered data (59 points in the present example)			
					Electrode intercept (mV)	Electrode slope (mV) per lg[H⁺]	Overall standard deviation (mV)	Number of residuals greater than 0.1 mV
1	13.890	A			480.5	59.23	0.3114	36
		B	0.16		480.4	59.24	0.2752	19
		C	1.7	1.5	479.9	59.16	0.2811	18
2	13.896	A			480.4	59.20	0.3395	30
		B	0.16		480.3	59.20	0.2763	18
		C	0.9	0.8	480.1	59.16	0.2798	18
3	13.902	A			480.4	59.16	0.3400	26
		B	0.16		480.3	59.17	0.2793	18
		C	0.26	0.1	480.2	59.16	0.2798	18
4	13.908	A			480.3	59.12	0.3408	23
		B	0.16		480.2	59.13	0.1818	18
		C	0.42	0.6	480.4	59.16	0.2798	18
5	13.914	A			480.2	59.08	0.3420	22
		B	0.16		480.1	59.09	0.2867	19
		C	1.1	1.3	480.6	59.16	0.2798	18
6	13.920	A			480.1	59.04	0.3436	20
		B	0.16		480.0	59.05	0.2909	19
		C	1.8	2.0	480.8	59.16	0.2798	18

* A = before CALIBT optimization; B = after optimization of [HCl]; C = after optimization of [NaOH]

Table 4.3. The essential results of MAGEC–MINIQUAD cycling on an oxalic acid–sodium hydroxide titration

Run no.	Program	E_{const} (mV)	lg β_{101}	lg β_{102}	Oxalic acid concentration (mmol dm^{-3})	MAGEC objective function	MINIQUAD R-factor
1	MAGEC	479.4 (OUT)	3.6 (IN)	4.7 (IN)	9.982 (IN)	0.3322	
2	MINIQUAD	479.4 (IN)	3.650 (OUT)	4.528 (OUT)	9.982 (IN)		0.003381
3	MAGEC	479.4 (OUT)	3.650 (IN)	4.528 (IN)	9.982 (IN)	0.04198	
4	MAGEC	479.3 (OUT)	3.648 (OUT)	4.510 (OUT)	9.982 (IN)	0.04125	
5	MINIQUAD	479.3 (IN)	3.648 (OUT)	4.497 (OUT)	9.982 (IN)		0.003379
6	MAGEC	479.5 (OUT)	3.648 (IN)	4.497 (IN)	9.942 (OUT)	0.03032	
7	MINIQUAD	479.5 (IN)	3.656 (OUT)	4.596 (OUT)	9.942 (IN)		0.003462
8	MAGEC	479.9 (OUT)	3.656 (IN)	4.596 (IN)	9.950 (OUT)	0.03351	
9	MINIQUAD	479.9 (IN)	3.662 (OUT)	4.688 (OUT)	9.950 (IN)		0.003449
10	MAGEC	479.9 (IN)	3.662 (IN)	4.688 (IN)	9.971 (OUT)	0.04418	
11	MINIQUAD	479.9 (IN)	3.660 (OUT)	4.670 (OUT)	9.971 (IN)		0.003406
12	MAGEC	480.0 (OUT)	3.660 (IN)	4.670 (IN)	9.971 (IN)	0.03681	

noticeably altered. In subsequent MAGEC runs, first the lg β's and later the ligand concentration were refined. (MINIQUAD, of course, can be used only to refine β values, not E_{const} or ligand concentration.) It may be seen from the table that lg β_{101} converged rapidly to within 0.01. E_{const} converged to within 0.1 mV and the eventual value obtained for the ligand concentration differed from the nominal value by a factor of 0.11%, which can be considered to be within experimental error. This example illustrates that the tactics used need to be judged at the beginning of each new run. The point at which one concludes the computations is also a matter of judgement and depends upon the criteria one uses. Taking goodness of fit as a criterion might persuade one to use the parameters obtained in run 5, whereas adopting convergence of the parameters as a criterion might take one as far as run 12 or perhaps even beyond this. One should, however, take cognizance of the significance of differences between successive MINIQUAD R-factors or successive MAGEC objective functions. Application of the Hamilton test (Hamilton 1964) to the R-factors may well reveal no significant differences between any of the MINIQUAD runs.

An important point is that experience enables a user to develop skill in applying the MAGEC–MINIQUAD cycling technique. The same can be said, of course, about the use of any computer program.

Other examples are listed in addition to a complete set of titration data in the MAGEC paper (May *et al.* 1982).

Discussion of MAGEC

Numerical analysis of titration data, to determine parameters such as E_{const}, would appear to have many advantages. In this respect, MAGEC is useful both (i) in the strong-acid–strong base titration range and (ii) in the hydrogen-ion-concentration range where the interaction of a given ligand with protons or a metal ion is significant. A strong advantage lies in the ability of MAGEC together with MAGEC–MINIQUAD cycling to determine electrode calibration parameters even when the protonation constants of the ligand are not known *a priori*. Thus, E_{const} is optimized for a given titration, and any change in the liquid-junction potential is incorporated into the particular E_{const} value obtained. Small variations in the liquid-junction potential arising, for example, from poor reproducibility or concentration effects, cease to be a problem, at least within the present state of the art.

A further marked advantage is derived from the MAGEC (CALIBT) analysis of strong-acid–strong-base data in assessing the reliability of a given glass electrode.

Yet another feature of the MAGEC (CALIBT) routine lies in the analytical facility whereby the concentration of a given strong-acid or strong-base solution can be determined more precisely than by the Gran method.

2. ACBA

General

ACBA (ACid-BAse titrations) refines any of the parameters in potentiometric acid–base titrations which may incorporate solutions containing mixtures of weak or strong acids or bases (Arena *et al.* 1979). It yields refined values of E_{const}, electrode slope, junction potentials, protonation constants of one or more ligands in the titrand and of ligand in the titrant, and concentrations of the components in the titration. Up to eleven titration parameters may be refined simultaneously. The program outputs values of the average degree of protonation (Z) for each ligand at each titration point.

Principles involved in ACBA

The symbolic coding for the ACBA program is listed in the Appendix. The computations involve solving the mass balance equations in the ligands and protons (cf. equations (4.4) and (4.5)) expressed in terms of the protonation constants for the ligands (cf. equation (4.3)). An additional equation (4.6) (applicable to each titration point) which is solved simultaneously with the mass balance equations is that expressing the electrode response (cf. equation (4.1)) but including the liquid-junction potential as an explicit term:

$$E = E'_{const} + s \lg [H^+] + j[H^+]. \tag{4.6}$$

The solutions are obtained by a non-linear least-squares iterative method (Arena et al. 1979) in which minimization is carried out of the error-squares sum of residuals in titre volume.

In the program, a distinction is made between parameters which are common to several titrations and parameters which have different values from titration to titration. The parameters in both groups are refined simultaneously.

Worked examples using ACBA (Arena et al. 1979)

Example 1
Two titrations of mixtures of acids and bases are reported in Table 4.4. The conditions were 25 °C and $I = 100$ mmol dm^{-3} NaClO$_4$.

Table 4.4. Acid–base titrations analysed by ACBA

	Taken	Found	Error
Mixture I*			
HClO$_4$ (mmol dm^{-3})	0.495	0.580	+17.2%
malonic acid (mmol dm^{-3})	1.452	1.422	−2.1%
succinic acid (mmol dm^{-3})	1.277	1.270	−0.5%
phthalic acid (mmol dm^{-3})	0.905	0.876	−3.2%
total acid (mmol dm^{-3})	7.763	7.716	−0.6%
malonic acid pK_1	2.623	2.66 ± 0.02	+0.04
pK_2	5.250	5.26 ± 0.02	+0.01
succinic acid pK_1	4.018	4.01 ± 0.01	−0.01
pK_2	5.147	5.08 ± 0.01	−0.07
phthalic acid pK_1	2.745	2.67 ± 0.01	−0.08
pK_2	4.920	5.06 ± 0.01	+0.14
Mixture II†			
pyridine (mmol dm^{-3})	1.543	1.549	+0.4%
2,2′-bipyridyl (mmol dm^{-3})	1.872	1.865	−0.4%
total base (mmol dm^{-3})	3.415	3.414	0.0%

* Two titrations; titrant 250.0 mmol dm^{-3} NaOH
† One titration; titrant 99.7 mmol dm^{-3} HClO$_4$

Example 2
A weak-acid–weak-base titration having titrand = acetic acid (15 cm^3, 606 mmol dm^{-3}) + NaClO$_4$ (95 cm^3, 160 mmol dm^{-3}); titrant = pyridine (5.081 mol dm^{-3}). Parameters refined are (i) E_{const}, (ii) titrant concentration, (iii) lg K_1 for acetate protonation, and (iv) lg K_2 for pyridine protonation. The sample of the input and output is given in Table 4.5.

Determination of hydrogen-ion concentrations 59

Table 4.5. ACBA analysis of weak-acid–weak-base titration

```
       Input
              PYRIDINE-CH3COOH  I=0.15(NACLO4)
              99 2 4E0    CTL2 K11   K21
              4.60
              5.30
              25.       -13.69
              PYRIDINE-CH3COOH  I=0.15(NACLO4)
              0.         0.        110.   440.    0.     0.
              0.60597    0.         1 0 0.13636
              0.         5.         0 0 0.

              0.21   227.0   00.50   201.8   00.80   185.9   01.00   177.8   01.20   170.6   0
              1.50   161.4   01.80   153.6   02.09   147.1   02.45   140.3   02.80   134.8   0
              3.32   128.2   03.90   122.2   04.40   117.9   04.85   114.5   1
Output
              PYRIDINE-CH3COOH  I=0.15(NACLO4)
LOG.K11=    4.600
LOG.K21=    5.300

 TEMPERATURE=   25.000
 SL=    59.159
 KWL=  -13.690

 TITRATION   1       PYRIDINE-CH3COOH  I=0.15(NACLO4)
    COH = 0.            FDILH= 0.             CTH = 0.            VO= 110.000     EO= 440.000     JA=   0.
    COL1= 0.60597000    FDIL1= 0.13636000     CTL1= 0.            NPO1= 1         NPT1= 0
    COL2= 0.            FDIL2= 0.             CTL2= 5.00000000    NPO2= 0         NPT2= 0

    NO. OF POINTS   14

TOTAL NO. OF POINTS   14

ST.DEV.= 0.414E 00     R(HAMILTON)= 0.133E 00       WITH THE INPUT DATA

CYCLE N.    4          ST.DEV.= 0.471E-02      R(HAMILTON)= 0.152E-02
                              VALUE                ST.DEV.
              EO         442.8560295          0.1335093E 00
              CTL2         5.0778247          0.4476654E-01
              K11          4.5174921          0.4945248E-02
              K21          5.2582256          0.2922945E-02

        V        DV         E        PH        Z1      Z2
  1   0.2100   0.0036    227.00    3.649     0.881   0.976
      0.5000   0.0037    201.80    4.075     0.735   0.938
      0.8000   0.0043    185.90    4.343     0.599   0.892
      1.0000  -0.0011    177.80    4.480     0.521   0.857
      1.2000  -0.0020    170.60    4.602     0.451   0.819
      1.5000  -0.0054    161.40    4.758     0.365   0.760
      1.8000  -0.0065    153.60    4.889     0.298   0.700
      2.0900  -0.0026    147.10    4.999     0.248   0.645
      2.4500   0.0037    140.30    5.114     0.202   0.582
      2.8000   0.0078    134.80    5.207     0.170   0.529
      3.3200   0.0027    128.20    5.319     0.136   0.465
      3.9000   0.0010    122.20    5.420     0.111   0.408
      4.4000  -0.0027    117.90    5.493     0.096   0.368
      4.8500  -0.0020    114.50    5.550     0.085   0.338
```

Example 3

Titration of a mixture of weak bases with a strong acid (Table 4.6). Titrand: 1.543 mmol dm^{-3} pyridine + 1.872 mmol dm^{-3} 2,2′-bipyridyl (input values 1 and 2). Titrant: 99.7 mmol dm^{-3} HClO$_4$. Parameters to be refined are (i) E'_{const}, (EO), (ii) concentration of pyridine, (iii) concentration of 2,2′-bipyridyl. In the output, Zi is equal to the average number of protons bound to the ligand i.

Table 4.6. ACBA analysis of the titration of weak bases with a strong acid

```
Input
            PYRIDINE-2,2-BIPYRIDYL I=0.1(NACLO4)
    99 2 3E0     COL1 COL2
    5.33
    4.461
    25.            -13.75
            PYRIDINE-2,2-BIPYRIDYL I=0.1(NACLO4)
    0.          0.0997    100.    420.    -480.    0.
    0.001       0.             0 0 0.
    0.002       0.             0 0 0.

    0.05    8.60    00.15   38.30   00.25   53.00   00.35   63.00   00.45   71.00   0
    0.60   80.60    00.75   88.90   00.90   96.10   01.04  102.2    01.15  106.7    0
    1.30  113.0     01.50  120.5    01.65  126.4    01.80  132.1    02.00  139.2    0
    2.15  144.9     02.30  150.5    02.46  156.9    02.65  164.4    02.81  171.6    0
    2.99  179.8     03.12  186.4    03.25  193.3    03.45  203.9    03.56  209.6    0
    3.70  216.1     03.85  221.9    04.00  227.2    04.15  231.8    04.40  238.0    1
```

Output

```
            PYRIDINE-2,2-BIPYRIDYL I=0.1(NACLO4)
LOG.K11=    5.330
LOG.K21=    4.461

    TEMPERATURE=    25.000
    SL=    59.159
    KWL=  -13.750

TITRATION  1        PYRIDINE-2,2-BIPYRIDYL I=0.1(NACLO4)
    COH = 0.          FDILH= 0.          CTH = 0.09970000   VO= 100.000    E0= 420.000    JA=-480.000
    COL1= 0.00100000  FDIL1= 1.00000000  CTL1= 0.           NPO1= 0        NPT1= 0
    COL2= 0.00200000  FDIL2= 1.00000000  CTL2= 0.           NPO2= 0        NPT2= 0

    NO. OF POINTS   30

TOTAL NO. OF POINTS   30

ST.DEV.= 0.456E 00      R(HAMILTON)= 0.175E 00    WITH THE INPUT DATA

CYCLE N.  6      ST.DEV.= 0.244E-02    R(HAMILTON)= 0.935E-03

                        VALUE              ST.DEV.
              E0      415.9951820        0.7832945E-01
              COL1      0.0015484        0.3855969E-05
              COL2      0.0018658        0.2824248E-05

        V          DV        E        PH       Z1     Z2
 1   0.0500-0.0010    8.60    6.886   0.027  0.004
     0.1500-0.0017   38.30    6.384   0.081  0.012
     0.2500-0.0006   53.00    6.136   0.135  0.021
     0.3500-0.0011   63.00    5.967   0.187  0.030
     0.4500-0.0000   71.00    5.832   0.240  0.041
     0.6000-0.0011   80.60    5.669   0.314  0.058
     0.7500 0.0020   88.90    5.529   0.387  0.079
     0.9000 0.0017   96.10    5.407   0.456  0.102
     1.0400-0.0006  102.20    5.304   0.515  0.125
     1.1500-0.0036  106.70    5.228   0.558  0.146
     1.3000 0.0025  113.00    5.122   0.618  0.179
     1.5000-0.0040  120.50    4.995   0.684  0.226
     1.6500 0.0023  126.40    4.895   0.731  0.269
     1.8000 0.0054  132.10    4.799   0.773  0.315
     2.0000-0.0029  139.20    4.679   0.818  0.377
     2.1500 0.0003  144.90    4.582   0.848  0.431
     2.3000-0.0016  150.50    4.488   0.874  0.485
     2.4600 0.0030  156.90    4.379   0.899  0.547
     2.6500-0.0030  164.40    4.252   0.923  0.618
     2.8100 0.0026  171.60    4.131   0.941  0.682
     2.9900-0.0026  179.80    3.992   0.956  0.747
     3.1200-0.0013  186.40    3.880   0.966  0.792
     3.2500-0.0002  193.30    3.763   0.974  0.833
     3.4500-0.0007  203.90    3.583   0.982  0.883
     3.5600 0.0023  209.60    3.486   0.986  0.904
     3.7000 0.0035  216.10    3.376   0.989  0.924
     3.8500-0.0032  221.90    3.277   0.991  0.939
     4.0000-0.0014  227.20    3.186   0.993  0.950
     4.1500 0.0018  231.80    3.107   0.994  0.958
     4.4000-0.0002  238.00    3.001   0.993  0.967
```

Discussion of ACBA

The examples demonstrate that agreement between known and calculated values is impressive. Further, internal calibration for E'_{const} is more than adequate without a separate calibration titration. Interestingly, the program has also been used to calibrate a copper-ion-selective electrode and the slope was found to be 29.7 ± 0.2 mV pCu^{-1} at $25.0\,°C$.

Special features of this program are that the refinements are carried out on the actual values of the parameters and not on their logarithmic values. (Such refinements are often ill-conditioned when logarithmic values are refined.) Because the parameters differ by many orders of magnitude, this dictates that the matrix of the coefficients has to be scaled accordingly. Under these conditions the ACBA technique for matrix inversion permits convergence of the iterative process even when only approximate estimates of the parameters are known.

3. LETAGROP

General

LETAGROP is a very versatile program that searches for the set of parameters that will minimize a certain function, typically an error-square sum (Arnek et al. 1969). It has been applied to many different problems, chemical and others.

The main chemical application has been to the refinement of formation constants. In this respect, the program has been described as the pioneering program of its type and several years ahead of its time (Hartley et al. 1980). The formation-constant refinement computations can be applied to potentiometric or spectrophotometric data. In the case of the former, Sillén realized early on the feasibility of refining electrode parameters and analytical concentrations simultaneously with formation constants (Brauner et al. 1969) thus enabling LETAGROP to be used as an electrode-calibrating program. This idea of simultaneously refining these parameters is one of the important principles used in MAGEC, ACBA and LIGEZ.

Principles involved in LETAGROP (Sillén 1962).

LETAGROP differs in its computing strategy from MAGEC, ACBA, LIGEZ and MINIPOT in that the functions which relate the parameters, known and unknown (e.g. equations (4.2)–(4.5)) need not be formulated explicitly in terms of the unknowns. Instead, the LETAGROP method involves seeking the functional behaviour of the error-square sum,

$$U = (a_{calc} - a_{obs})^2$$

by a process called pit-mapping. The user has the choice of expressing the error-square sum in terms of any of the measured variables such as emf or titre.

First, the user supplies a set of initial estimates of the unknown parameters which are used to calculate a value of U. Subsequently, a series of adjustments is made to the unknown parameter values and a value of U is calculated corresponding to each combination of the latter. Regions of progressively lower and lower U values are sought, thereby enabling the U-surface to be defined.

With two unknown parameters, the U-surface takes the form of an elliptical paraboloid or pit. The pit is formed in N-dimensional space when the number of unknown exceeds 2. The object of defining the pit is to find the bottom, i.e. the minimum value of U. When the minimum is reached as closely as possible, the corresponding set of unknown parameters is taken as the 'best' set. In practice, the minimization is carried out by fitting the pit to an algebraic expression (using a Taylor series).

In the context of the present chapter, the E_{const} value for the glass/reference-electrode cell would be determined in this way, thereby utilizing LETAGROP as an electrode calibration program.

Examples using LETAGROP

Many worked examples are to be found in the earlier literature, but nowadays there are more recently developed algorithms and programs available which converge more rapidly from poorer estimates of the parameters than can be refined by LETAGROP and so examples are not listed in this book.

4. MINIPOT

General

MINIPOT is one of the important programs which has been developed specifically for running on a Wang 2200 desk computer; 16 kbytes of core memory are needed (Gaizer and Puskas 1981). The program comprises two distinct parts, the first providing a method for calibrating a glass/reference-electrode cell using data obtained in a strong-acid–strong-base titration. The electrode-parameter values so obtained may then be used in the second part of the program for computing formation-constant values of up to four binary mononuclear complexes. The latter may be either ligand–proton complexes or metal–ligand complexes.

Principles involved in the MINIPOT algorithm

The first part of the program is more pertinent to the present chapter than the second. The latter, of course, is more relevant to Chapter 5 of this book.

In the first part, the data (titre, emf) in the concentration-buffered regions from a strong-acid–strong-base titration are processed. These are used to

Determination of hydrogen-ion concentrations

optimize the parameters, E_{const}, s, j'_H and j'_{OH} in the electrode equation, (4.7).

$$E = E_{const} + s \lg[H^+] - j'_H[H^+] - j'_{OH}K_W/[H^+]. \qquad (4.7)$$

In addition, the user may choose to optimize the concentration of the titrant and the value of K_W applicable to water or any other appropriate solvent.

The second part of the program processes data from a suitable potentiometric titration, making use of the parameter values obtained in part 1. This yields values of

$$\beta_{qp} = [B_q L_p]/[B]^q[L]^p$$

being the formation constant for a reaction

$$qB + pL = B_q L_p.$$

Here, B may be either proton or a metal ion. A maximum of four complexes may be treated.

Worked examples using MINIPOT (Gaizer and Puskas 1981)

The literature publication describing MINIPOT gives a detailed account of two examples of the use of MINIPOT; the first is a strong-acid–strong-base titration in propylene glycol–water mixture. Values of K_W (for the solvent concerned), j'_H, j'_{OH}, alkali concentration, E_{const} and s are obtained. The second example involves titrating a polypeptide solution with strong base and defining the parameters listed above with the exception of the alkali concentration. As this is a new program which could have widespread use on desk computers, both examples and the program listing are quoted in full in the Appendix.

Discussion of MINIPOT

Unlike ACBA, in the interests of reducing the running times, refinements are effected in terms of the logarithms of the values of the parameters in question other than E_{const} and s. This means that matrix elements are of comparable magnitudes, which simplifies the mathematics and thus curtails running times. The price to be paid, however, is that relatively good starting values are required. The 16-kbyte capacity of the desk computer used limits the calculation to about forty points. Similar programs have been used by Gaizer and Puskas for refining other experimental data such as those from spectrographic measurements.

5. LIGEZ

General

One of us (R.G.T.) has developed another program for use with a desk calculator. LIGEZ (*LIG*and *EZ*ero) can be run on an extended Hewlett

Packard (HP41C) Calculator fitted with a Quadram memory module and a Maths Pac module. The program written for the calculator can process up to 60 data points obtained from the titration of a monoprotic or diprotic weak acid with a strong base. It will simultaneously refine the electrode parameters, E_{const} and s, the protonation constants of the conjugate base of the weak acid being titrated, and some of the reagent concentrations.

Principles involved in LIGEZ

LIGEZ has been tested mainly on simulated titration data and on real data obtained for the titration of acetic acid with sodium hydroxide. As LIGEZ has not appeared in the literature, a full description together with the coding for a HP41C calculator is given in the Appendix. The mass balance equations (in the total dissociable protons and total ligand) are combined to yield a polynomial in $[H^+]$, the free hydrogen concentration. If the parameters of the weak-base–strong-acid system, such as the concentrations of the reagents, the protonation constants and the ionization constant of water, are known, this equation can be solved at each point using Newton's reiterative method (Conte and de Boer 1972). If the parameters are not known, trial values are used to give estimated values of $[H^+]$ and these, together with trial values of the electrode parameters, give estimates of E, the measured emf. The refinement procedure detailed in the Appendix then leads to corrections which improve the trial values. The system is considered to have converged when changes in the parameters being refined are less than the precision that can be expected from the data.

Worked examples using LIGEZ

Example 1

A hypothetical set of data was created representing the titration of a mixture of strong acid and a weak acid such as glycine, which has one displaceable proton (NDP = 1), but which can in strong acid solution add a further proton.

The data were generated by giving values to the acid–base system parameters and by assuming that the equation obeyed by the electrode system was

$$E = E_{const} + s \ln [H^+]. \qquad (4.8)$$

The values given to the various parameters were: $V_{init} = 20$ cm^3, $K_W = 2.088 \times 10^{-14}$, $\beta_{101} = 4.186 \times 10^9$, $\beta_{102} = 1.0659 \times 10^{12}$, $[LH]_{init} = 5.29$ mmol dm^{-3}, $[MA]_{init} = 5.31$ mmol dm^{-3}, $[SB] = 19.42$ mmol dm^{-3}, $s = 25.69$, $E_{const} = 440.31$ mV. The acid–base parameters were used in the polynomial in $[H^+]$ already mentioned to obtain values of $[H^+]$ which were then used in equation (4.8) to obtain values of E. These simulated values of E calculated for various values of V, the volume of strong base added, are tabulated in Table D1 in Appendix D. In using these data to test LIGEZ, the following procedure was adopted.

Determination of hydrogen-ion concentrations

(i) The concentrations of the strong acid and strong base were assumed to be known and were not refined.

(ii) The following parameters were estimated as shown below and were refined:

$s = 25.0$, $K_W = 1 \times 10^{-14}$ mol² dm⁻⁶, $\beta_{101} = 1 \times 10^9$, $\beta_{102} = 1 \times 10^{11}$, $[LH]_{init} = 5.4$ mmol dm⁻³, $E_{const} = 400.000$ mV.

The refinement to convergence took six cycles and this is set out in Table 4.7.

The last column labelled $\sum (E_{obs} - E_{calc})^2$ is a measure of the goodness of fit of the calculated emf (E_{calc}) with the observed emf (E_{obs}). The last row of the table lists the 'true' values of the parameters, i.e. those used to simulate the titration. The rather large value obtained, viz. 0.2982 mV², for $\sum (E_{obs} - E_{calc})^2$, when these true values were used is due to round-off errors which were made and deliberately left in when the simulated data were manufactured. Each refinement cycle takes about 43 minutes to process 51 data points.

Example 2

A titration of acetic acid (NDP = 1) with sodium hydroxide in 1.00 mol dm⁻³ sodium chloride was performed. Application of LIGEZ to the experimental data in the region before the endpoint is shown in Table 4.8. The concentration of the sodium hydroxide was 19.46 mmol dm⁻³ and was not refined. In the region of the titration curve being used the calculated hydrogen-ion concentration is insensitive to K_W and hence K_W was arbitrarily set at 2×10^{-14} and was not refined. V_{init}, the volume of acetic acid titrated, was 20.00 cm³. The actual data are tabulated in Table D2 in Appendix D and the four refinement cycles needed to obtain convergence are set out in Table 4.8. Each cycle ran for about 28 minutes and 46 points were processed.

Discussion of LIGEZ

The method underlying LIGEZ is very well known and of general application, and electrode equations other than that used in the examples cited above can be applied with only small modifications to the program. For monoprotic or diprotic acids and strong bases a calculator, although slow, can be used. For more complex systems a microcomputer can be employed. Convergence, provided reasonable estimates of the parameters are started with, is rapid, usually less than six cycles. It does happen that if values of the protonation constants started with are too large they will in the course of a refinement swing negative, resulting in the failure of the next cycle. In such a case a new, smaller, estimate of the parameter must be used. Generally it is better to start with trial values which are too low rather than too high. Failure will also occur if the starting value of [H⁺] used in solving the equation for [H⁺] is lower than the solution being sought. Hence, the initial value of [H⁺] which is input into the program must always be higher than it is actually likely to

Table 4.7. LIGEZ refinement cycles for the simulated titration of a glycine-like substance in strong acid with a strong base

Cycle number		s (mV/ln[H$^+$])	$K_W \times 10^{14}$	$\beta_{101} \times 10^{-9}$	$\beta_{102} \times 10^{-11}$	[LH]$_{init} \times 10^3$ (mol dm^{-3})	E_{const} (mV)	$\sum (E_{obs} - E_{calc})^2$ (mV^2)
				Parameters to be refined				
1	INPUT	25.000	1.0000	1.0000	1.0000	5.400	400.000	22592.8
	REFINEMENT	1.102	−0.3506	1.3870	2.5981	0.408	43.652	
2	INPUT	26.102	0.6494	2.3870	3.5981	5.808	443.652	3511.4
	REFINEMENT	−0.054	1.2935	1.0181	3.0067	−0.132	−1.472	
3	INPUT	26.048	1.9429	3.4051	6.6048	5.676	442.180	170.1
	REFINEMENT	−0.350	0.3786	0.7055	3.3531	−0.417	−1.625	
4	INPUT	25.698	2.3215	4.1106	9.9579	5.529	440.555	29.879
	REFINEMENT	−0.012	−0.2586	0.0967	0.7597	0.031	−0.284	
5	INPUT	25.686	2.0629	4.2073	10.7176	5.290	440.271	0.064166
	REFINEMENT	−0.003	0.0136	0.0039	0.0385	−0.004	−0.017	
6	INPUT	25.683	2.0765	4.2112	10.7561	5.286	440.254	0.044253
	REFINEMENT	—	—	—	—	—	—	
True values		25.690	2.0890	4.1860	10.6590	5.290	440.310	0.2982

Determination of hydrogen-ion concentrations

Table 4.8. LIGEZ refinement cycles for the titration of acetic acid with sodium hydroxide in 1.00 mol dm^{-3} sodium chloride at 25 °C

Cycle number		s (mV/ln[H$^+$])	$\beta_{101} \times 10^{-4}$	[LH]$_{init} \times 10^3$ (mol dm^{-3})	E_{const} (mV)	$\Sigma (E_{obs} - E_{calc})^2$ (mV^2)
			Parameters to be refined			
1	INPUT	25.690	1.0000	10.000	420.000	19499.56
	REFINEMENT	1.047	1.3170	−0.213	1.472	
2	INPUT	26.737	2.3170	9.787	421.472	4164.48
	REFINEMENT	−0.841	0.6250	0.128	7.217	
3	INPUT	25.896	2.9420	9.915	428.689	62.99
	REFINEMENT	0.080	0.1209	0.024	0.483	
4	INPUT	25.976	3.0629	9.939	429.172	0.1547
	REFINEMENT	0.002	0.0012	—	0.012	
5	INPUT	25.978	3.0641	9.939	429.184	0.1362
	REFINEMENT	—	—	—	—	

The experimental determination of formation constants

Clearly, the precisions with which one achieves electrode-calibration and formation-constant measurements are closely related. There is, however, another factor—the choice of experimental equipment and techniques. This topic has recently been reviewed in a IUPAC publication which describes internationally agreed guidelines for the determination of stability constants (Nancollas and Tomson 1982). The article is aimed at a wide range of scientists, especially those whose background is not in physical chemistry but who need reliable constants applicable to their own conditions and special ligands. Researchers wishing to get started in this field are strongly advised to use the approaches described by the IUPAC Commission.

Concluding remarks and our experiences

In aqueous solution it is generally true that a ligand which complexes with a metal ion over a range of pH also becomes protonated over the same pH range. In simple terms

$$L+H \rightleftharpoons LH \quad \beta_{101}$$

$$L+M \rightleftharpoons LM \quad \beta_{110}.$$

Since ion-selective electrodes for determining either [L] or [M] are not sufficiently reliable to work below 10^{-5} mol dm^{-3}, the use of glass electrodes and the fact that [L] is a common link between these two equations are used to quantify β_{110}; i.e. β_{101} is known, [H] is measured, [L] thus calculated and β_{110} computed.

Clearly, the real monitoring system is, in practice, a combination of a glass electrode and a pK (or β_{101}) titration. It thus seems particularly apposite that glass/reference-electrode calibration *in situ*, measured during a pK titration, and covering the pH range over which the ligand L becomes protonated, ought to be used as the best available means of calibration for metal-complexing work.

Furthermore, as E_{const} values sometimes vary with the concentration of ligand present, although not in a reproducible or predictable manner, it is prudent to conduct the ligand-protonation–electrode-calibration experiment at the same ligand concentration as the metal-complexing ligand titrations. The ideal situation is one of calibration *in situ* (Rossotti 1978). The closer one approaches this objective, the more accurate are one's results.

Determination of hydrogen-ion concentrations

The following general procedure is recommended:

(i) Record a 'buffer-line' (see Chapter 2) to establish electrode reliability.
(ii) Perform a strong-acid–strong-base titration and subject data to MAGEC analysis as the CALIBT routine can produce further evidence of the soundness of quality of the electrode.
(iii) A series of acidified ligand solutions are titrated with base throughout the pH range and processed by an approach such as MAGEC-MINIQUAD cycling to yield electrode parameters and the ligand pK values.
(iv) Metal-ligand complexing titrations are performed at ligand concentrations similar to those used in (iii).

Clearly, the further one proceeds with this scheme, the larger the number of parameters one could vary. The more these are varied, the greater the effort required in the experimental and computational calibration procedures. Accordingly, it is convenient to arrange parameters into three broad categories:

(i) Those which need determining for each experimental run, for example, E_{const}, β_{LMH}.
(ii) Those which are determined less frequently, e.g. pK_W, ligand pK's, and analytical concentrations.
(iii) Those which are taken from the literature, e.g. hydrolysis constants for the metal ions.

By a judicious distribution of parameters among these three categories much time and effort can be saved and a better cognizance of the reliability of one's results obtained. This is important because the speciation models computed in the next chapter can only be as reliable as the constants obtained in this chapter and used as input data for speciation models.

It is noteworthy that the procedure described above using MAGEC-MINIQUAD cycles and CALIBT could equally well be written in terms of ACBA and LETAGROP, etc. One might wonder whether discussions at future solution-chemistry conferences will centre upon which programs and procedures mentioned in this chapter are more fundamental, and to be accepted as 'standards' just as the dichotomies concerning whether electrodes ought to be a_{H^+} calibrated by one, or by several, standard buffers have been debated in the past (see Chapter 3). Again, it is important to realize that there is far more that unites than divides supporters of these assorted approaches, since all are now able to calibrate electrodes in terms of [H$^+$] to within a fraction of a millivolt, and to produce pure concentration formation constants rather than mixed constants. Both of these are tremendous improvements of which researchers can be justly proud.

Many of us have faced the choice of writing our own computer programs—the only really efficient means of knowing exactly what a program is doing—or of accepting well-tried software using programs written by other researchers for tackling solution problems related to, but differing in some details from, the research project in hand. The consistent advice of computer scientists concerning this common quandary is always to opt for the established software, even if it means expense in terms of purchasing programs from a supplier, in terms of staff training to use such programs, and in terms of small modifications necessitated by the handling of the program.

The alternative approach, whereby one creates new programs, is usually an extremely expensive option in terms of man-hours and cpu time. The small gain accrued costs a lot of effort. There are parallels in terms of microprocessor control of the new generation of 'intelligent instruments' for the analytical laboratory. Typically, the microprocessor programming of a new instrument costs up to £5000 000 and takes about 10 man-years, thus accounting for up to 50 per cent of development costs (Ford 1982). The decision to commence writing a new program ought to be well thought out and priced before being considered.

Three final pieces of advice are to be commended—first, whenever a difficult decision has to be taken fairly subjectively (for example, whether to choose the 'best' set of constants, in titres, in residuals, R factors, etc.), the eventual compromise decision can always benefit from the experience of rereading the wisdom of the Sillén school's publications of the 1950s and 1960s. Secondly, we ought never to ignore good experimental evidence. For example, rather than computationally adjust an E_{const} value that appears to have jumped between duplicate titrations, just compare the initial emf of both to see whether it too has jumped. These simple steps can often suggest pipetting errors or burette leaks. Thirdly, care must be exercised when parameters calculated using programs such as LIGEZ are to be judged in terms of their accuracy. This is illustrated by the first example given in the section describing LIGEZ. All experimental data contains random error. In the case of the example cited above, this random error was introduced into the simulated data by rounding-off errors when the emf was calculated from the 'true' parameters; the rather large value of $0.2982\ \text{mV}^2$ obtained for $\sum (E_{obs} - E_{calc})^2$ underlines this. The values obtained for the parameters using the program differ from the 'true' value as can be seen by comparing the figures in the bottom two lines of Table, 4.7, but if judged in terms of the value they give for $\sum (E_{obs} - E_{calc})^2$, they give a better fit to the data. Thus, it is always wise, where possible, to determine parameters, such as protonation constants, under a number of different experimental conditions and then to assess their accuracy and independence of random and rounding off errors in terms of constancy of the output parameters.

Were it not for these difficult decisions, research into new areas of chemistry would lack a lot of its lure and excitement.

5

Determination of complex concentrations and speciation

The previous chapter has frequently made reference to the concentrations of each of the species present during a calibration titration or in an electrode cell. Such concentrations can be computed relatively easily from a knowledge of the formation constants for all interactions that might occur, and are produced as a routine printout when programs are being used to refine formation constants or to calibrate electrode pairs. Programs which produce such complex concentrations include LETAGROP (Ingri and Sillén 1964), MINIQUAD (Sabatini *et al.* 1974), MAGEC (May *et al.* 1982), SUPERQUAD (Gans *et al.* 1983), and many others. Often the [H^+] is varied as part of a titration and concentration versus $-\lg[H^+]$ plots are produced.

However, researchers are interested in the effects of other parameters being varied as well as pH and so programs such as COMICS (Perrin and Sayce 1967), SCOGS, (Sayce 1968), HALTAFALL (Ingri *et al.* 1967), and ECCLES (May *et al.* 1977), were specially composed to perform these calculations.

Input data required for a speciation calculation

The total concentration value for each component is required unless its free-ion concentration is known. For example, for the blood plasma models each amino-acid was entered as a total concentration, whereas the free-proton concentration and *free*-metal-ion concentrations were specified instead of their totals (May and Williams 1977).

All relevant formation constants must be used as input. These must include the possibility, no matter how unlikely, of every ligand present complexing with every metal ion in a variety of hydrolysed and protonated states. Solubility products, partition coefficients, partial pressure constants and, of course, solution formation constants are all extracted from the literature, or measured, and entered as data. This means very many constants are listed, for example, the blood plasma model commonly uses a data-bank of more than 10 000 complexes (May *et al.* 1977; Berthon *et al.* 1978).

There are now hundreds of thousands of formation constants available from the literature. These have been listed in special publications (Sillén and Martell 1964, 1971; Perrin 1979) critically surveyed by scientists who are experts in respect of certain metal–ligand systems (Smith and Martell 1976; Anderegg 1977; Beck 1977; Högfeldt 1982). Thus, the user has the choice of (i) finding exactly the constant and conditions in the literature appropriate to the speciation model proposed, (ii) using a recommended constant obtained by a critical assessment of the literature, or (iii) measuring a constant using the guidelines suggested at the end of the previous chapter. Such constants are now widely used in elucidating mechanisms of reactions and designing buffers, and in speciation studies.

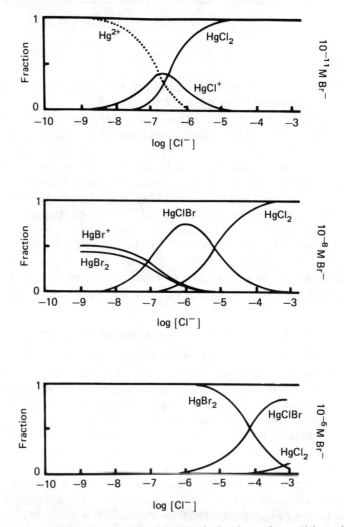

Figure 5.1 A speciation plot showing the optimization of conditions for preparing an elusive compound such as HgClBr (Dyrssen *et al.* 1968). Clearly, both the free chloride and bromide concentrations need to be optimized simultaneously

To attain this last objective, these formation constants and concentration data are used as input for a series of computer programs which progressively estimate equilibrium concentrations until all mass balance equations are satisfied to within a specified tolerance. The output may be plotted (as in the examples shown in Figures 5.1 and 5.2) or given as a listing of printed columns.

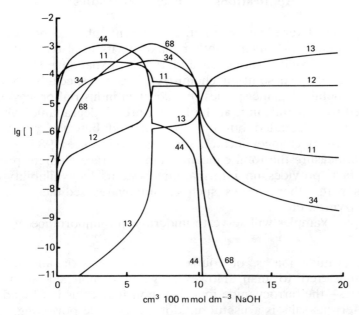

Figure 5.2 Speciation during a complex hydrolysis titration of $Pb(NO_3)_2$ (50.0 cm^3, 10.0 mmol dm^{-3}) with NaOH (100 mmol dm^{-3}) (Dyrssen *et al.* 1968). The numbers indicate the different Pb_mOH_n complexes present

Nordstrom *et al.* (1979b) report a comparative review of a variety of recently developed speciation programs. With each they have computed the species distribution in two hypothetical test samples, one of river water and the second of sea water.

The programs were found to yield major species concentrations in the river water which agreed excellently with each other. Less impressive agreement amongst the major species concentrations was found in the sea water. In both test samples, the minor species tended to show serious discrepancies in their concentrations as computed by the different programs. The authors attribute all the discrepancies to several factors and cite the most important as differences in the thermodynamic data bases built in to certain of the programs.

It follows that very careful judgement is needed in selecting a program to be used for speciation computations. Preferably, one should use more than one program in a given project. Furthermore, it is clearly necessary to be deeply critical of the data-base used for the calculations, particularly if the program is of the variety that has a built-in data-base. Finally, it must be emphasized that the results obtained by Nordstrom *et al.* underscore the vital need for continuing effort among experimentalists to determine meticulously the formation constants for complexes in single-metal-ion–single-ligand systems.

Applications of speciation studies

Speciation, the determination of the different physicochemical species formed by an element which together make up its total concentration in a sample, is now a well established and rapidly expanding topic for many sciences which involve solution chemistry. This is especially so when the element is a metal. Organic species *in vivo* or in industrial processes may be distinguished from metals in that the former are biodegradable, whereas the latter cannot be degraded but may be changed between poisonous and non-toxic forms by changes in their speciation (Florence 1982).

Merely to measure the total concentration of a trace element present in a sample, in fact, provides no information about its bioavailability or of its interactions with other species such as soil water sediments, suspended particles, enzymes, etc.

Two simple examples will serve to underline the importance of speciation knowledge.

(i) The cyanide ion is, of course, exceedingly toxic and lethal when administered to man. However, when complexed as a different species—the ferrocyanide ion—it is non-toxic; this is why administering ferrous salts is a useful antidote to cyanide poisoning.

(ii) The potential toxicity of a metal, even though it is not biodegradable, can nevertheless be completely controlled by a change in its speciation. Mercury metal is relatively harmless and is widely used in dentistry and in temperature-measuring instruments in the clinic, whereas organic forms of mercury are lethal poisons and are to be excluded from man's environment at all cost.

Much of the value of speciation knowledge centres upon the general concept of bioavailability. It is a well-established dogma that a metal, which to all intents and purposes is essential for human life, can be rendered completely non-bioavailable and useless by changing its speciation. One may quote the total precipitation of iron from our diet by certain dietary components. At the other end of the spectrum, one finds that a metal which is normally present in only infinitesimally small amounts, but which is regarded as beneficial to man, for example, nickel, can be rendered completely bioavailable in absolute terms to such an extent that it becomes highly toxic; for example, nickel salts can be converted into nickel carbonyl which rapidly penetrates cell membranes and is, not surprisingly, highly toxic therein.

Over the last few years, an increasing number of inroads have been made into studies of speciation in all walks of life. Methods used for speciation include anodic-stripping voltammetry, ion exchange, ultrafiltration, dialysis, and ion-selective electrode measurements, but this last method is usually too insensitive for the measurement of heavy metal ion concentrations in naturally occurring samples.

A technique in which there have been major advances made over the last decade is that of computer modelling of trace-metal speciation (May and Williams 1977; May et al. 1977; Berthon et al. 1978; Huang et al. 1982; Williams 1984). It is based upon the concept that many solutions in nature or in industry are in a steady state and, in order to maintain efficiency, must therefore be pretty close to thermodynamic equilibrium. It provides an attractive alternative to the difficult and tedious experimental techniques listed in the previous paragraph. In its crudest form it is able to produce models which set limits on speciation; in the more refined applications of the subject it has impressively produced detailed speciation versus biological response correlations.

The centrepiece of the technique is a data-bank of all conceivable formation constants, preferably in concentration rather than activity terms, applicable to the process under scrutiny. It is inevitable in the near future that much of the legislation concerning effluents from industry and the maintenance of water quality will, in fact, include statements relating to speciation. Similarly, submissions to the Federal Drug Authority and the Committee on Safety of Medicines in future will need to contain sections concerning the trace-element effects of a new drug and speciation evidence will play a central role in this topic. Further examples will be given in subsequent paragraphs.

Unfortunately, as the steady state is a kind of quasi-equilibrium situation, the concept of lability must be taken into consideration. Many of the species present are there at extremely low concentrations which are not analysable by any known straightforward method. However, when one tries to concentrate these labile species one completely disturbs the thermodynamic equilibrium and thus the species which eventually predominates or which precipitates from solution may bear very little resemblance to that which occurred under the conditions of study. Ion-selective electrodes and the glass electrode are able to monitor solutions without disturbing this equilibrium because they function through very low currents and have very high internal resistances. Similarly, if a computer model is constructed to reflect the equilibria present it in no way disturbs the equilibrium being investigated; this seems to be an ideal means of studying speciation. However, it is necessary to have reliable concentration-formation constants for each and every possible reaction which may occur and this is only made possible through precise glass-electrode potentiometry and analysis in terms of millivolts and proton concentrations. Details of such models are given in the following sections.

Complex-formation research

A knowledge of exact speciation composition and concentrations (as distinct from activities) of ions present in solution can enable one to optimize conditions for preparing and isolating a rare complex. A simple example is given in Figure 5.1 (Dyrssen *et al.* 1968).

Furthermore, a knowledge of concentration speciation arising from calibrations of glass electrodes in terms of proton concentrations can give a much deeper understanding of the principles underlying analytical techniques and permit a reinforcement of the rigour of the theories involved. For example, in Figure 5.2 (Dyrssen *et al.* 1968) an example of a sophisticated competing complexing titration is given. It may be seen that such a graph, calculated using a model of the titration based upon concentration-formation constants, can indicate the choice of ion-selective electrode which could well be used to monitor the progress of this titration, the free-ion concentration ranges through which this electrode is expected to function, and, where one to feed solubility products into this model, it could reveal the areas of the titration where precipitation is likely to cause interference.

Clearly, in principle it would be highly desirable to have all analytical procedures quantified in terms of speciation plots similar to those shown in Figures 5.1 and 5.2.

Medicine

One may use glass electrodes and hydrogen-ion-concentration data either to study the pathology of a disease or to study the influence upon trace-metal concentrations of treating the disease using pharmaceuticals.

Conditions which prevent the synthesis of a key biological component (for example, the lack of ceruloplasmin in Wilson's Disease which overloads other copper-complexing ligands) can completely upset the normal steady-state equilibrium prevailing in a biological fluid. The magnitude of this imbalance and an indication of which species is likely to precipitate first or to change its composition may be calculated from formation constants. For example, a neutral, bioavailable species may well become a charged, non-bioavailable species when an additional ligand is added to the central metal ion (Halstead and Williams 1983).

An area in which the growth of research of late can only be described as explosive is that of the influence of administered drugs upon trace-metal concentrations. There are many trace metals which occur *in vivo* in concentrations higher than those desirable for normal human biochemistry, and so excessive amounts need to be removed using chelating ligand therapy. Ligand drugs may be chosen such that they have the right donor groups to be somewhat selective (no ligand drug is absolutely specific) for the offending metal ion and then speciation studies will determine whether the drug will be capable of winning the metal ion from literally thousands of other competing reactions which occur in the biofluid under study. The removal of such metal ions is often a multistage process in that one ligand is necessary in order to take the metal from the cell into blood plasma (the species must of course be neutral in this respect) and then another ligand is used in order to form a charged complex with the metal so that the complex is then suitable for renal excretion. The competitiveness of these two ligands and the competition from all other naturally occurring ligands is one that can only be assessed using models. As previously mentioned, such simulation models now extend to more than 10 000 low-molecular-weight complexes occurring in biological fluids such as blood plasma (May and Williams 1977; May et al. 1977; Berthon et al. 1978).

Similarly, industrial metals such as cadmium, nickel, lead, or plutonium, may be removed by the judicious use of ligand drugs; once again, synergistic chelation therapy whereby two ligands are used in order to chaperone the metal through various biological barriers is usually the most effective. Examples are given in Table 5.1.

Table 5.1 Synergistic chelation therapy. A few examples of how different ligand drugs chaperone metal ions from one biofluid into another as part of the therapy of excess metal ion removal. Formulae of the drugs listed are shown in Figure 5.3

Metal ion	Agent promoting movement from tissue/cells into plasma	Agent promoting movement from plasma into urine
Copper	D-penicillamine	D-penicillamine Triethylenetetramine N,N'-bis(2-aminoethyl)-1,3-propanediamine
Nickel	Sodium diethyldithio-carbamate	Tetraethylenepentamine
Lead	D-penicillamine, ICRF 226 British Anti-Lewisite	Calcium, sodium edetate

It has now been realized that many of the traditional drugs used for treating a whole range of conditions from high blood pressure to tuberculosis or arthritis are, in fact, capable of complexing with metal ions found *in vivo*. Such metal-dependent side effects of drug therapy must now be assessed before submissions are placed to the legislative bodies which permit the marketing of a new agent and in some respects it is often necessary to top-up the patient with essential trace metals inadvertently removed by the therapy. The selection of 'lead' metals for further investigation in respect of side effects can best be produced using computer simulation approaches, since it is not possible to screen all pharmaceuticals and their influences upon all biometals under all conditions. Furthermore, such simulation models will permit an optimization of therapy such that the maximum desirable therapeutic effect is achieved whilst minimizing the trace-element imbalances caused by the drug. Examples are given in Table 5.2.

Table 5.2 Examples of pharmaceuticals which unintentionally form trace metal complexes *in vivo*. Formulae are shown in Figure 5.4

Agent	Treatment for	Trace-element complexes probably formed	Result	Reference
Ethambutol (Lederle Labs Ltd)	Tuberculosis	Metabolic product wherein 2-$CH_2OH \rightarrow 2COO^-$ complexes zinc in plasma	Zinc depletion side effects noted	Cole *et al.* (1981)
Prizidilol (SK & F Ltd)	Hypertension	4-nitrogen ring+side-chain $\rightarrow$ 8-nitrogen dimer by autoxidation and ferrous complex	Red complex formed at high drug concentrations in some animals	Al-Falahi *et al.* (1984)
Razoxane (ICI Ltd)	Cancers	Metabolic product is ICRF 198 which complexes a range of transition metal ions	Cytotoxicity of this and and its homologues related to degree of zinc complexing and lipophilicity	Huang *et al.* (1982)
Captopril (E.R. Squibb Ltd)	Hypertension	Copper and zinc complexes formed *in vivo*	Some side-effects similar to D-penicillamine	Hughes and Williams (1984)

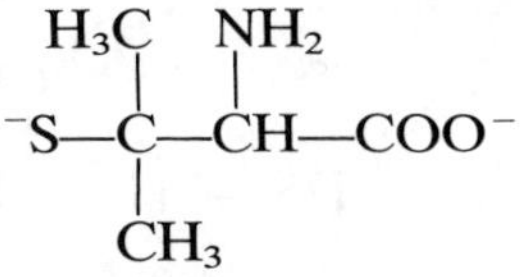

Figure 5.3 Ligand drugs mentioned in Table 5.1

$$^-S-\underset{\underset{CH_3}{|}}{\overset{\overset{H_3C}{|}}{C}}-\underset{\underset{}{}}{\overset{\overset{NH_2}{|}}{CH}}-COO^-$$
D-Penicillamine

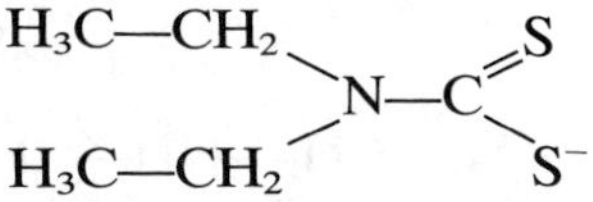

$$\begin{matrix}H_3C-CH_2\\ \diagdown\\ H_3C-CH_2\diagup\end{matrix}N-C\begin{matrix}\diagup\!\!\!\!^S\\ \diagdown S^-\end{matrix}$$
Diethyldithiocarbamate

$H_2N-CH_2-CH_2-NH-CH_2-CH_2-NH-CH_2-CH_2-NH_2$
Triethylenetetramine

$H_2N-CH_2-CH_2-NH-CH_2-CH_2-CH_2-NH-CH_2-CH_2-NH_2$
N,N'-bis(2-aminoethyl)-1,3-propanediamine

$H_2N-CH_2-CH_2-NH-CH_2-CH_2-NH-CH_2-CH_2-NH-CH_2-CH_2-NH_2$
Tetraethylenepentamine

$(^-OOC-CH_2)_2N-CH_2-CH_2-N(CH_2-COO^-)_2$
Edetate

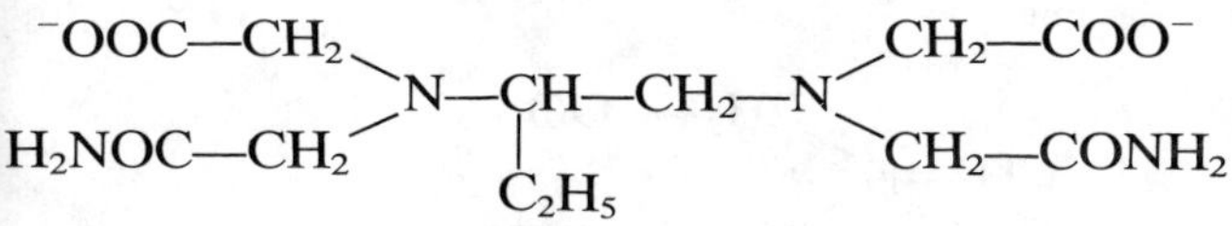

$$\begin{matrix}^-OOC-CH_2\\ \diagdown\\ H_2NOC-CH_2\diagup\end{matrix}N-\underset{\underset{C_2H_5}{|}}{CH}-CH_2-N\begin{matrix}\diagup CH_2-COO^-\\ \diagdown CH_2-CONH_2\end{matrix}$$
ICRF 226

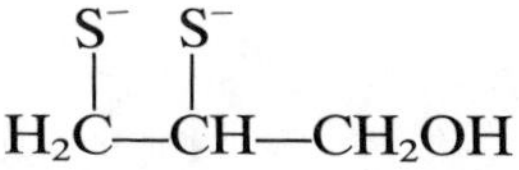

$$H_2\underset{\underset{}{}}{\overset{\overset{S^-}{|}}{C}}-\underset{\underset{}{}}{\overset{\overset{S^-}{|}}{CH}}-CH_2OH$$
British Anti-Lewisite

Industrial uses

Industry has progressed through various levels of sophistication since the industrial revolution. Increasing competitiveness in terms of prices and greater vigilance by health and safety bodies has resulted in far greater monitoring of industrial reactions. It is no longer sufficiently cost-effective or environmentally acceptable merely to judge what goes on in the reaction vessel from a knowledge of total amounts of input and of output respectively. On the contrary, it is now highly desirable that we are able to predict the likely outcome of a reagent impurity, of a deficiency in one reactant, of localized overheating, and indeed of very many parameters which may well vary in the real world of industry. Such objectives demand a detailed speciation knowledge. This will permit adequate safety margins and shut-down procedures to be built into the industrial process so that the reaction vessel is shut down and rendered safe long before the occurrence of irreparable damage, or the outlet of toxic materials. It is, of course, rather naive to believe that all industrial processes are determined solely by formation constants: a detailed knowledge of mixing efficiencies, viscosities, flow characteristics, etc. is also very necessary. However, having set up as

C_2H_5
|
CH—NH—CH$_2$—CH$_2$—NH—CH
| |
CH$_2$OH   CH$_2$OH
         Ethambutol

           CH$_3$ COO$^-$
            | |
HS—CH$_2$—CH—C—N
              ‖
              O
              Captopril

[pyridazine-phenyl]—NHNH$_2$
with N=N
OCH$_2$—CH(OH)—CH$_2$—NH—C(CH$_3$)$_3$
         Prizidilol

HN[diketopiperazine]N—CH—CH$_2$—N[diketopiperazine]NH
                      |
                      CH$_3$
         Razoxane

$^-$OOC—CH$_2$ \   / CH$_2$—COO$^-$
               N—CH—CH$_2$—N
H$_2$NOC—CH$_2$ / | \ CH$_2$—CONH$_2$
                     CH$_3$
         ICRF 198

Figure 5.4 Ligand drugs mentioned in Table 5.2

detailed a model as possible with all such constants and characteristics as input, one may compare the model's output with species concentrations as monitored by glass or other ion-selective electrodes.

Ideally, one should have this speciation knowledge at each stage of the complete industrial process; but, in practice, one tends to have a limited amount of knowledge of certain micro areas of the process which have deserved a greater amount of research and study. Eventually, one hopes that a macro speciation model of all modern industrial processes will be available. It is possible to calculate buffer capacities, not only of certain micro regions, thus indicating the degree of resilience to slight imperfections in the input material, but also for the whole macro system.

A global example may be of use here. Many governments now have a finance model based upon the cost of raw materials such as oil, coal, hydroelectricity, etc., from many countries of the world. When the model works well it is capable of predicting the increase in price of electricity in one country arising from a variation in the cost per barrel of oil from another country, such oil passing through yet more countries. The oil may never actually be used to generate electricity in the country under study but the economic repercussions of financial fluctuations in raw material prices have ripples which spread out across most of the globe. One hopes that speciation models of the chemical industry will be somewhat more precise and predictable than such global economic models and that they will result in our chemical industry becoming even more reliable, more safe, and more cost-effective in the future.

Returning to glass-electrode monitoring and formation constants being based upon concentrations of protons rather than activities, we must emphasize the fact that industry does not work at the desirable thermodynamic concentration of almost infinite dilution but often uses thick pastes of extremely concentrated chemicals being mixed and reacted together. Such industrial concentrations ought still to obey formation constants based upon concentration scales but it will be exceedingly difficult, and probably impossible, to calculate activity coefficients for these sludges.

An interesting example of the application of a speciation study to the resolution of an industrial problem has been reported by Allison et al. (1975). The problem is concerned with the separation of the zinc-containing mineral, sphalerite, from the co-occurring constituents of the mined ore, particularly chalcopyrite (a copper sulphide mineral), pyrite, galena, mica, silica and calcite. The separation is effected by a flotation process in which all the ore-constituents except sphalerite are floated to the surface of the liquid medium whereas the sphalerite is depressed to the bottom.

Cyanide ions are generally added to de-activate the sphalerite particle surfaces against reaction with xanthate collectors used to promote the flotation of the other constituents. Zinc sulphate is added in order to enhance depression of the sphalerite particles.

At the Prieska Copper Mine the problem took the form of poor metallurgical performance and high reagent consumption. In an endeavour to improve the efficiency of the flotation process, the investigation referred to by Allison et al. (1975) was undertaken with the aim of elucidating the mechanism of sphalerite depression. Included in the investigation was an extensive speciation study of the aqueous zinc(II) ion–cyanide anion system. Ranges of total concentrations of these components, zinc:cyanide ratios and pH were covered (Harris and Marsicano 1975). The effects of carbonate anions and calcium(II) ions were also examined (Marsicano et al. 1976).

One of the numerous distribution diagrams constructed by these researchers is reproduced in Figure 5.5. While the distribution pattern varies somewhat from one set of conditions to another, Figure 5.5 is typically representative of the general features of the system. The speciation results combined with results obtained in other types of experiment lead to the conclusion that it is the solid precipitates and not the dissolved species which bring about depression of the sphalerite particles.

The effective precipitates are basic zinc sulphate, zinc hydroxide and zinc carbonate. The mechanism appears to be adsorption of the precipitate on to the sphalerite surface, thus rendering the latter hydrophilic and hence tending to reduce or eliminate any natural floatability of the mineral arising from hydrophobicity of the surface. Although the important question of selectivity of depression in mixed sulphide systems remains unanswered,

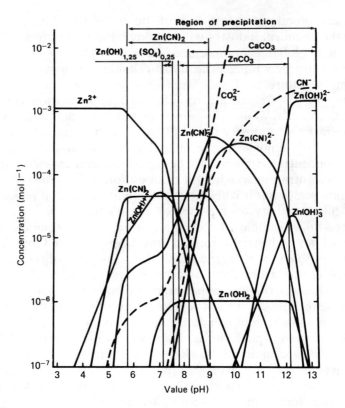

Figure 5.5 Distribution of soluble species in a slurry having total zinc = total sulphate = 1.47, total cyanide = 2.24 and total calcium = 1.17 mmol dm^{-3} at equilibrium with atmospheric carbon dioxide

these speciation studies have made important contributions towards a more rapid and meaningful optimization of reagent additions and pH-selection in the operation of this flotation plant.

The oceans and the atmosphere

The illustrious Lars Gunnar Sillén paved the way to applying formation constants (and solubility products) to the chemistry of the oceans and atmosphere (Sillén 1967). He based his treatment on the idea of geochemical balance (Goldschmidt 1933) whereby the oceans, ocean sediments and the atmosphere were formed over many millions of years by chemical reactions between igneous rocks and volatiles (H_2O, HCl, CO_2, compounds of S, B, N, F, etc.), the latter being the same as are still ejected by volcanoes. These reactions may be represented as in (5.1):

$$\text{igneous rock} + \text{volatiles} \rightarrow \text{sea water} + \text{sediments} + \text{air.} \qquad (5.1)$$

Sillén proposed an imaginary experiment in which a sea water–sediments–air mixture with constituent composition as estimated by Goldschmidt (and revised by Horn and Adams (1966)) is allowed to reach equilibrium. He used Gibbs' phase rule in his arguments and introduced a hypothesis of heterogeneous equilibria between cations and aluminosilicate minerals, for example,

$$1.5Al_2Si_2O_5(OH)_4(s) + K^+ \rightleftharpoons KAl_3Si_3O_{10}(OH)_2(s) + 1.5H_2O + H^+.$$
$$\text{kaolinite} \qquad\qquad\qquad \text{K-mica}$$

(5.2)

These considerations, remarkably, enabled him to account for the pH and main cationic concentrations, provided the temperature and chloride concentration are regarded as fixed. The ocean, of course, is not at equilibrium but is, in fact, a dynamic system which is being subjected continuously to a complex network of input and output material fluxes.

Sillén considered the possibility of reaction (5.1) being a two-way process still in progress, as an alternative to its being a one-way process which had taken place when the oceans were formed. Furthermore, he noted that there is a continual interaction between the sea and the rest of the earth's crust: the ocean evaporates, producing rain which attacks rocks and soil, resulting in dissolved ions and suspended solids that the rivers deposit in the sea. The suspended solids can interact with seawater by ion exchange or phase transformations releasing hydrogen ions and other cations. The ionic balance can be further upset by underwater volcanism, life processes, metamorphic changes, dust and the uprising and submergence of land masses.

One of the principal conclusions reached by Sillén was that in spite of the continuous input and removal of material, the pH and main cationic concentrations are maintained more or less constant by heterogeneous equilibria such as (5.2). Provided no phase is allowed to disappear completely, the system acts as a pH-stat (rather than a buffer). This is at strong variance with the earlier view that the ocean pH is buffered by the $CO_3^{2-} + HCO_3^- + H_2CO_3$ system!

Following on from Sillén's treatment, Goldberg (1963) listed the principal chemical species occurring in sea water and pointed out the likelihood of ion-pair formation between sulphate and divalent metal ions. Subsequently, Dyrssen and Wedborg (1974) improved the input to HALTAFALL by using an expanded list of formation constants and computed the speciation of elements in a standard sea water composition. The list of complexes likely to be formed in the sea was expanded further by Ahrland (1975). Whitfield (1975) proposes a greatly improved approach to seawater speciation through combining his model of specific interionic interaction with the equilibrium model.

Sillén's concept of the regulation of the sea's composition by heterogeneous equilibria between sea water and oceanic sediments (over geological time) is regarded in many quarters as being seminal (for example, Stumm and Morgan (1981), McDuff and Morel (1980)). Nevertheless, there are critics. Ahrland (1975), for example, suggests that a direct contact between the bottom sediments and any small volume of ocean water occurs too infrequently (about once in 1600 years) for this explanation of pH control to be truly satisfactory.

Current views adopt the idea that the concentrations of most of the elements in the ocean are fixed by a dynamic balance between input and removal processes (McDuff and Morel 1980; Stumm and Morgan 1981; Whitfield and Jagner 1981). The authors just mentioned do point out, however, that the dynamic and equilibrium concepts complement each other. McDuff and Morel (1980), in particular, highlight the pertinence of Sillén's concept of metamorphic processes between minerals in explaining the 'alkalinity' value of the oceans.

Environmental problems

Nordstrom et al. (1979a) have used the program WATEQ2 (Ball et al. 1979), to investigate the speciation of iron(II) and iron(III) in acid mine waters and stream water contaminated by drainage of the former. The dominant solute iron complexes, together with the precipitates formed at low pH at the mine sites and subsequently in the flowing streams where the pH increases, have been identified. These computations, in conjunction with appropriate field observations improve the insight into the chemical processes of acid mine waters, the redox relations of iron and solution–mineral reactions.

Mattigod and Sposito (1979) demonstrate the application of the program, GEOCHEM, to trace metal equilibria in a mixture of irrigation water and geothermal brine and also to the aqueous phase of a sewage-sludge-amended soil.

Bioavailability of metals as toxins or nutrients

Magnusson et al. (1979) report the application of the program, REDEQL2, to calculation of speciation of copper(II) in a preliminary factor analysis study of the toxicity of copper towards *Daphnia magna* (the water flea). Their tentative conclusions suggest the copper(II) aquo ion, neutral and cationic hydroxo complexes, to be the most toxic. The anionic copper hydroxo complexes appeared to be significantly less toxic, whereas the carbonato copper complexes showed no toxicity. Although these authors state that more extensive work is in progress and is yet to be published, the present results may be considered as paradigmatic in illustrating how bioavailability of metals is dictated by speciation.

While yet other examples are to be found in the literature, the final application we mention in this chapter is of models that have been developed by Lindsay and his co-workers (Lindsay 1979) for examining the role of natural and synthetic chelating agents in supplying nutrients to plant roots. Several types of chelating agents have been considered in relation to a range

Determination of complex concentrations and speciation

of metal nutrients under a wide variety of conditions. The general conclusion reached is that the chelating agents tend to increase the level of nutrient cations in solution thereby increasing availability by increasing both mass-flow and diffusion of the nutrients to the roots.

Lessons learned, and their application to other ion-selective electrodes

Glass electrodes are no longer just another physical chemical probe. During this last decade, there has been a renaissance in their uses and to many research scientists they have become a way of life. They are used not only to monitor, but to protect both patients in hospitals and processes in industry, and also to control important reactions.

We have tried to indicate in this book that there are various levels of sophistication at which a glass electrode may be used and we have tried to indicate the appropriate level to be matched to a given use. We also hope that we have drawn attention to the levels of sophistication on either side of the choice made in order to permit the scientist to take his results much further with a little more effort placed into calibration and perhaps understanding the theory.

This has all become possible because glass electrodes are now able, given the right techniques, to produce numbers which are both stable and reproducible and so it is necessary for glass electrode users to catch up in terms of mastering the theories of calibration and monitoring. The pay-off, if one takes the time and effort to calibrate in terms of hydrogen-ion concentration rather than the activity of the hydrogen ion, is one of vast new vistas in terms of speciation. Such speciation models of virtually any solution which exists in nature are not available to those who calibrate electrodes in terms of pH.

It is pleasing to see that ion-selective-electrode researchers have always preferred to calibrate in terms of metal-ion concentrations, usually using a millivolt scale. This serendipity has arisen because of the difficulty in deriving buffers for metal-ion concentrations. Thus, not surprisingly, ion-selective-electrode monitoring of metal-ion concentrations can be a useful adjunct to glass electrode potentiometry and one hopes that ion-selective electrodes will be developed to greater sensitivities and selectivities over the next few years. One area in which ion-selective electrodes may well be improved is that of the least-squares assessment of series of data points, from a titration or from individual measurements. The objective approaches built into such large hydrogen-ion-concentration programs such as MAGEC could well be applied with good result to ion-selective-electrode work.

Concluding remarks

Most analytical chemists prefer to know the concentrations present in their samples rather than activities. Unfortunately, many fluids of interest have species too dilute to analyse and so computer simulation is necessary. It must be stressed, however, that whenever possible, models and experiments—involving biological samples, etc.—must go hand in hand. This can mean designing *in vivo* experiments to produce 'spot' calibrations of the computer models; such experiments may, perhaps, involve spiking the biofluid until some species is analysable and comparing model theory with experiment.

Any speciation conclusion based upon a computer model is only as good as the input data. Despite three decades of extensive effort directed towards obtaining unequivocal formation constants for metal–ligand systems, there remain worrying uncertainties concerning the accurate calibration of the electrodes and the precise measurement of a_{H^+} and $[H^+]$—the latter in particular. It usually helps to clarify the situation if one discusses these problems. This book has aimed to do just that; at best it could lead to a new era of speciation analysis. At the very least, it might have clarified some of the problems by discussing them in terms of concentration and speciation.

APPENDICES*
Computer programs referred to in the text

Appendix A. The MAGEC program (May et al., 1982)

The following data input is required for each titration.

Item	Flag	Description
1		The title
2	PKW	Ionization constant of water (as negative lg)
3	PK1	First stepwise protonation constant (as lg)
3	PK2	-up to six acid pK values can be accepted
4	VZO	Initial volume in vessel
5	H+V	Mineral-acid concentration in vessel
6	LGV	Ligand concentration in vessel
7	H+B	Mineral concentration in burette
8	LGB	Ligand concentration in burette
9	EZO	Glass electrode intercept parameter
10	SLP	Glass electrode slope parameter
11		Titration data pairs (vol, emf). Up to 100 points.

Format requirements

ITEM 1:   (20A4)
ITEMS 2–10:   (A3, 2X, I1, 1X, I1, 3G10.3)
ITEM 11:   (2G10.3)

Data requirements

ITEMS 2–10:   A3 -Flag as defined above
                    I1 -Refined key

---

* The MAGEC, ACBA and MINIPOT programs, and two MINIPOT examples, are reproduced by kind permission of the copyright owners, Pergamon Press Ltd.

	I1 -Ligand NDP (only for ITEMS 6 and 8)
	G10.3-Value or estimate of parameter
	G10.3-Lower refinement limit of value
	G10.3-Upper refinement limit of value

Refine keys: 0 = no refinement
                1 = normal refinement
                2 = refine between upper and lower limits
NDP: Number of dissociable protons on the ligand
Mineral acid: Use negative values for alkali concentrations

Program listing

```
C     ******************************************************************
C
C
C
C                         PROGRAM MAGEC.
C                         -------------
C
C
C
C
C                         MULTIPLE ANALYSIS
C                         OF TITRATION DATA FOR
C                         GLASS ELECTRODE CALIBRATION.
C
C
C
C     ******************************************************************
C
C
C
C     THE PROGRAM ANALYSES POTENTIOMETRIC DATA FROM TITRATIONS IN WHICH
C     THE TITRAND AND THE TITRANT ARE (1) A STRONG ACID OR A STRONG
C     BASE, (2) A LIGAND OR (3) A COMBINATION OF (1) AND (2).
C
C     ESTIMATES OF THE CONCENTRATIONS, THE EQUILIBRIUM CONSTANTS AND
C     THE PARAMETERS FOR THE NERNSTIAN EQUATION OF THE GLASS ELECTRODE
C     RESPONSE TO HYDROGEN ION CONCENTRATION CAN BE REFINED.  THE USER
C     SPECIFIES WHICH ARE TO BE SIMULTANEOUSLY OPTIMISED.  UPPER AND
C     LOWER LIMITS CAN BE IMPOSED ON ALL PARAMETERS BEING REFINED.
C
C     THE OPTIMISATION IS PERFORMED BY SUBROUTINE NELM AND GENERALLY,
C     THE EQUILIBRIUM CONCENTRATIONS ARE EVALUATED BY SUBROUTINE ML AND
C     SUBROUTINE FUNCT.
C
C     SUBPROGRAM CALIBT ONLY ACCEPTS DATA PERTAINING TO MONOBASIC
C     REACTANTS.   IT IS MAINLY USED TO ANALYSE STRONG ACID VERSUS
C     STRONG BASE TITRATIONS.   THE PARAMETERS OF THE NERNSTIAN
C     EQUATION ARE FOUND BY LINEAR LEAST SQUARES BEST FIT.
C     IN ADDITION IT PERFORMS GRAN PLOT CALCULATIONS, SCANS VALUES OF
C     PKW AND PK1 AND ADJUSTS THE CONCENTRATIONS OF REACTANTS IF THIS
C     IMPROVES THE AGREEMENT BETWEEN THE THEORETICAL AND THE OBSERVED
C     SLOPE FOR THE GLASS ELECTRODE RESPONSE.
C
C
C     THE PROGRAM WAS DEVELOPED AT THE UNIVERSITY OF WALES INSTITUTE
C     OF SCIENCE AND TECHNOLOGY IN 1978. IT IS WRITTEN IN FORTRAN IV.
C
C
C     ******************************************************************
```

Appendix A: MAGEC

```
C
C
C
      INTEGER TITLE, LIT(15)
      DIMENSION X(7), Y(7), H(7), XS(7), FP(8), ID(7)
      INTEGER OUT, REFINE
      REAL LVESSL, LBURET
      LOGICAL TITEND
      COMMON /ONE/  PK(7,3), VZERO(3), HVESSL(3), LVESSL(3), HBURET(3),
     *  LBURET(3), EZERO(3), SLOPE(3), REFINE(15), NDPV, NDPB
      COMMON /TWO/  CI(7),CX(2),TT(2),HX(2),TOLC(2),DT(2),DDT(2,2)
      COMMON /THREE/ V(100), E(100), NP
      COMMON /FOUR/ BETA(7), ARRAY(3), TOL, NCONST, NBETAH, JQR(2,7)
      COMMON /FIVE/  IN, OUT, IFAIL, JFAIL, AL10, KOUNT
      COMMON /SIX/   TITLE(20), P(63)
C
C
      DATA   LIT /'PKW','PK1','PK2','PK3','PK4','PK5','PK6',
     *    'VZO','H+V','LGV','H+B','LGB','EZO','SLP','    '/
C
C
C
10200 FORMAT(20A4)
10231 FORMAT(8G10.3)
10205 FORMAT(A3,2X,I1,1X,I1,2X,3G10.3)
20000 FORMAT('1')
20001 FORMAT(///'0',1H*,3X,1H*,6X,3(1H*),7X,4(1H*),5X,5(1H*),6X,
     *   4(1H*),/' ',2(2(1H*),1X),4X,1H*,3X,1H*,5X,3(1H*,9X),
     *   /' ',1H*,1X,1H*,1X,1H*,5X,5(1H*),5X,1H*,1X,3(1H*),
     *   5X,4(1H*),6X,1H*,/' ',3(1H*,3X,1H*,5X),1H*,9X,1H*,
     *   /' ',2(1H*,3X,1H*,5X),1X,4(1H*),5X,5(1H*),6X,4(1H*),
     *   /' ',45(1H-),////)
20200 FORMAT(////'0',30X,20A4)
20203 FORMAT(////'0',4X,'INPUT VALUES FOR THE TITRATION PARAMETERS ARE:'
     *    ,///'0',4X,'IDENTIFIER',4X,'REFINE KEY',7X,'NDP',10X,'VALUE',
     *    26X,'LOWER AND UPPER LIMITS',//)
20204 FORMAT('0',8X,A3,12X,I1,13X,F17.3,17X,2F15.3)
20205 FORMAT('0',8X,A3,12X,I1,13X,1PE17.4,17X,2E15.4)
20206 FORMAT('0',8X,A3,12X,I1,12X,I1,1PE17.4,20X,2E15.4)
20208 FORMAT('+',89X,'N/A')
20234 FORMAT(///'0DIMENSION LIMITS OF TITRATION DATA ARRAYS EXCEEDED')
20250 FORMAT(///'0','SEQUENCE ERROR DETECTED IN THE DATA')
20260 FORMAT(///'0','ERROR IN THE LIMITS FOR REFINEMENT')
20270 FORMAT(///'0','ACID DISSOCIATION CONSTANT ERROR')
20290 FORMAT('0','EXECUTION TERMINATED.',///)
20302 FORMAT(///'0',4X,'V(OBS.)  V(CALC.)  EMF(OBS.)  EMF(CALC.)',
     2   7X,'RESIDUALS',12X,'PH',9X,'FL',8X,
     2   'EZO(CALC.)   SLP(CALC.)',///)
20305 FORMAT(////'0',2X,'REFINED VALUES FOR THE PARAMETERS ARE NOW:',///
     1   '0',2X,'OBJ. FUNCT.',5X,A4,6(8X,A4))
20306 FORMAT('+',102X,'OPTIMIZATION RECORD',//)
20310 FORMAT('0',1P8E12.3)
20311 FORMAT('+',102X,7(1H*),3X,'CONVERGED',//)
C
C
C     SECTION ONE:   INITIALISATION.
C
C
      IN = 8
      OUT = 12
      CALL FACSF('@ASG,UP 12. . ')
      DEFINE FILE 12(APRNTA,,132)
      IFAIL = 0
      JFAIL = 0
      EPS = 1.0E-6
```

```
              TOL = 1.0E-4
              AL10 = ALOG(10.00)
              TITEND = .FALSE.
C
C
C     SECTION TWO:    DATA INPUT
C
C
  200 WRITE(OUT,20000)
      READ(IN,10200,END=299)  TITLE
      WRITE(OUT,20001)
      WRITE(OUT,20200)   TITLE
      WRITE(OUT,20203)
      M = 1
      DO 207 I=1,7
      READ(IN,10205)  KEY, IVAL, JVAL, (PK(I,J), J=1,3)
      N = 1
      IF(IVAL.EQ.2)  N = 3
      WRITE(OUT,20204)   KEY, IVAL, (PK(I,J), J=1,N)
      IF(N.EQ.1)  WRITE(OUT,20208)
      IF(KEY.EQ.LIT(8))   GO TO 210
      IF(KEY.NE.LIT(I))   GO TO 250
      DO 205 L=1,3
  205 ARRAY(L) = PK(I,L)
      IF(IVAL.GT.0)   CALL SETUP(ID,LIT(I),M,X,H,IVAL,ARRAY)
      IF(I.LE.2)  GO TO 207
      IF(PK(I,1).LT.PK(I-1,1))   GO TO 270
  207 REFINE(I+7) = IVAL
      READ(IN,10205)  KEY, IVAL, JVAL, VZERO
      IF(KEY.NE.LIT(8))  GO TO 290
      IF(IVAL.EQ.2.AND.(PK(I,2).GT.PK(I,1).OR.PK(I,3).LT.PK(I,1)))
     *    GO TO 260
      GO TO 215
C
C     REFINE(1)    -   VZERO
C     REFINE(2)    -   HVESSL
C     REFINE(3)    -   LVESSL
C     REFINE(4)    -   HBURET
C     REFINE(5)    -   LBURET
C     REFINE(6)    -   EZERO
C     REFINE(7)    -   SLOPE
C     REFINE(8)    -   PKW
C     REFINE(9) TO REFINE(14)    -   PK1 TO PK6
C
  210 NCONST = I - 1
      DO 212 J=1,3
  212 VZERO(J) = PK(I,J)
  215 REFINE(1) = IVAL
      IF(REFINE(1).EQ.2.AND.(VZERO(2).GT.VZERO(1).OR.VZERO(3)
     *    .LT.VZERO(1)))   GO TO 260
      IF(REFINE(1).GT.0)   CALL SETUP(ID,LIT(8),M,X,H,REFINE(1),VZERO)
      READ(IN,10205)  KEY, REFINE(2), JVAL, HVESSL
      N = 1
      IF(REFINE(2).EQ.2)   N = 3
      WRITE(OUT,20205)   KEY, REFINE(2), (HVESSL(J), J=1,N)
      IF(N.EQ.1)  WRITE(OUT,20208)
      IF(KEY.NE.LIT(9))   GO TO 250
      IF(REFINE(2).EQ.2.AND.(HVESSL(2).GT.HVESSL(1).OR.HVESSL(3)
     *    .LT.HVESSL(1)))   GO TO 260
      IF(REFINE(2).GT.0)   CALL SETUP(ID,LIT(9),M,X,H,REFINE(2),HVESSL)
      READ(IN,10205)  KEY, REFINE(3), NDPV, LVESSL
      N = 1
      IF(REFINE(3).EQ.2)   N = 3
      WRITE(OUT,20206)   KEY, REFINE(3), NDPV, (LVESSL(J), J=1,N)
      IF(N.EQ.1)  WRITE(OUT,20208)
```

Appendix A: MAGEC

```
      IF(KEY.NE.LIT(10))    GO TO 250
      IF(REFINE(3).EQ.2.AND.(LVESSL(2).GT.LVESSL(1).OR.LVESSL(3)
     *   .LT.LVESSL(1)))   GO TO 260
      IF(REFINE(3).GT.0)    CALL SETUP(ID,LIT(10),M,X,H,REFINE(3),LVESSL)
      READ(IN,10205) KEY, REFINE(4), JVAL, HBURET
      N = 1
      IF(REFINE(4).EQ.2)   N = 3
      WRITE(OUT,20205)   KEY, REFINE(4), (HBURET(J), J=1,N)
      IF(N.EQ.1)   WRITE(OUT,20208)
      IF(KEY.NE.LIT(11))    GO TO 250
      IF(REFINE(4).EQ.2.AND.(HBURET(2).GT.HBURET(1).OR.HBURET(3)
     *   .LT.HBURET(1)))   GO TO 260
      IF(REFINE(4).GT.0)    CALL SETUP(ID,LIT(11),M,X,H,REFINE(4),HBURET)
      READ(IN,10205) KEY, REFINE(5), NDPB, LBURET
      N = 1
      IF(REFINE(5).EQ.2)   N = 3
      WRITE(OUT,20206)   KEY, REFINE(5), NDPB, (LBURET(J), J=1,N)
      IF(N.EQ.1)   WRITE(OUT,20208)
      IF(KEY.NE.LIT(12))    GO TO 250
      IF(REFINE(5).EQ.2.AND.(LBURET(2).GT.LBURET(1).OR.LBURET(3)
     *   .LT.LBURET(1)))   GO TO 260
      IF(REFINE(5).GT.0)    CALL SETUP(ID,LIT(12),M,X,H,REFINE(5),LBURET)
      READ(IN,10205) KEY, REFINE(6), JVAL, EZERO
      N = 1
      IF(REFINE(6).EQ.2)   N = 3
      WRITE(OUT,20204)   KEY, REFINE(6), (EZERO(J), J=1,N)
      IF(N.EQ.1)   WRITE(OUT,20208)
      IF(KEY.NE.LIT(13))    GO TO 250
      IF(REFINE(6).EQ.2.AND.(EZERO(2).GT.EZERO(1).OR.EZERO(3)
     *   .LT.EZERO(1)))   GO TO 260
      IF(REFINE(6).GT.0)    CALL SETUP(ID,LIT(13),M,X,H,REFINE(6),EZERO)
      READ(IN,10205) KEY, REFINE(7), JVAL, SLOPE
      N = 1
      IF(REFINE(7).EQ.2)   N = 3
      WRITE(OUT,20204)   KEY, REFINE(7), (SLOPE(J), J=1,N)
      IF(N.EQ.1)   WRITE(OUT,20208)
      IF(KEY.NE.LIT(14))    GO TO 250
      IF(REFINE(7).EQ.2.AND.(SLOPE(2).GT.SLOPE(1).OR.SLOPE(3)
     *   .LT.SLOPE(1)))   GO TO 260
      IF(REFINE(7).GT.0)    CALL SETUP(ID,LIT(14),M,X,H,REFINE(7),SLOPE)
C
      DO 231 I=1,100
      READ(IN,10231,END=235)   V(I), E(I)
      IF(V(I).LT.0.000)   GO TO 239
  231 CONTINUE
      I = 101
      READ(IN,10205,END=235)   KEY
      WRITE(OUT,20234)
      GO TO 290
  235 TITEND = .TRUE.
  239 NP = I - 1
      GO TO 300
C
  250 WRITE(OUT,20250)
      GO TO 290
  260 WRITE(OUT,20260)
      GO TO 290
  270 WRITE(OUT,20270)
  290 WRITE(OUT,20290)
  299 STOP
C
C
```

```
C       SECTION THREE:     REFINEMENT BY SUBROUTINE NELM.
C
C
    300 NBETAH = NCONST - 1
        M = M - 1
        IF(NBETAH.LT.1)   GO TO 400
        DO 301 I=1,NBETAH
        JQR(1,I) = I
    301 JQR(2,I) = 1
        JQR(1,NCONST) = -1
        JQR(2,NCONST) = 0
C       IF(M-1)   350, 320, 340
C   320 KOUNT = -50
C       X(1) = Y(1) + H(1)
C       WRITE(OUT,20302)
C       CALL CALC(M,X,F,&325)
C   325 X(1) = Y(1) - H(1)
C       WRITE(OUT,20302)
C       CALL CALC(M,X,F,&327)
C   327 X(1) = Y(1)
C       GO TO 350
        IF(M.EQ.0)   GO TO 350
    340 KOUNT = -25
        WRITE(OUT,20302)
        CALL CALC(0,X,F,&341)
    341 KOUNT = 0
        WRITE(OUT,20305)   (ID(I), I=1,M)
        WRITE(OUT,20306)
        DO 344 I=1,M
    344 Y(I) = X(I)
        IF(M.LT.7)   Y(M+1) = SLOPE(1)
        M1 = M + 1
        MP = M * (M + 2)
        CALL NELM(X,F,EPS,P,FP,XS,H,0,M,M1,MP)
        IF(KOUNT.LE.M)   GO TO 355
        WRITE(OUT,20310)   F, (X(I), I=1,M)
        IF(IFAIL.EQ.0)       WRITE(OUT,20311)
        IFAIL = 0
C
    350 KOUNT = -100
        WRITE(OUT,20302)
        CALL CALC(M,X,F,&355)
C       RESET ORIGINAL PARAMETER VALUES.
    355 REFINE(7) = 2
        SLOPE(3) = SLOPE(1) - ABS(SLOPE(1)) / 10.00
        SLOPE(2) = SLOPE(1)
    360 CALL CALC(M,Y,F,&400)
C
C
C       SECTION FOUR
C
C
    400 IF(NBETAH.GT.1.OR.ABS(LBURET(1)).GT.1.0E-8)   GO TO 999
        IF(LVESSL(1).GT.1.0E-8.AND.HBURET(1).LE.0..AND.NDPV.NE.1) GOTO 999
        IF(LVESSL(1).GT.1.0E-8.AND.HBURET(1).GT.0..AND.NDPV.NE.0) GOTO 999
        PKWMIN = PK(1,1)
        PKWMAX = 0.000
        BLGMIN = PK(2,1)
        BLGMAX = 0.000
        IF(REFINE(8).NE.2)   GO TO 401
        PKWMIN = PK(1,2)
        PKWMAX = PK(1,3)
        GO TO 450
    401 IF(REFINE(9).NE.2)   GO TO 450
        BLGMIN = PK(2,2)
        BLGMAX = PK(2,3)
```

Appendix A: MAGEC

```
  450 CALL CALIBT(VZERO(1),EZERO(1),SLOPE(1),HVESSL(1),LVESSL(1),
C
     *    HBURET(1),PKWMIN,PKWMAX,BLGMIN,BLGMAX)
C
C
  999 IF(.NOT.TITEND)   GO TO 200
      STOP
      END
      SUBROUTINE SETUP(ID,LIT,M,X,H,IREF,ARRAY)
      DIMENSION  X(7), H(7), ARRAY(3), ID(7)
      INTEGER OUT
      COMMON /FIVE/  IN, OUT, IFAIL, JFAIL, AL10, KOUNT
      IF(M.GE.8)   GO TO 20
      IF(IREF.EQ.2)   GO TO 10
      IXA = 1
      XA = ARRAY(1) / 12.000
      IF(XA.LT.0.0000)   IXA = -1
      ARRAY(2) = ARRAY(1) - XA * IXA
      ARRAY(3) = ARRAY(1) + XA * IXA
   10 H(M) = (ARRAY(2) + (ARRAY(3) - ARRAY(2)) / 10.0) - ARRAY(1)
      X(M) = ARRAY(1)
      ID(M) = LIT
      M = M + 1
      RETURN
   20 WRITE(OUT,99)
   99 FORMAT(//'0','ATTEMPT TO REFINE TOO MANY PARAMETERS',/' ',
     1 'ERROR TERMINATION',///)
      STOP
      END
      SUBROUTINE NELM(X,F,EPS,P,FP,XS,H,IX,N,N1,NP)
      DIMENSION  X(N), P(NP), FP(N1), H(N), XS(N)
      ISH = 1
      IS = 0
      NN = N * (N + 1)
      CALL CALC(N,X,F,&100)
      FP(1) = F
      IF(IX.NE.0)   GO TO 2
      DO 1 I=1,N
      K = I
      DO 1 J=1,N1
      P(K) = X(I)
      IF(I-J+1.NE.0)   GO TO 1
      P(K) = X(I) + H(I)
    1 K = K + N
    2 K = 1 + N
      DO 3 I=2,N1
      DO 4 J=1,N
      X(J) = P(K)
    4 K = K + 1
      CALL CALC(N,X,F,&100)
    3 FP(I) = F
      IF(FP(1)-FP(2).GT.0.0)   GO TO 5
      IH = 2
      IL = 1
      GO TO 6
    5 IH = 1
      IL = 2
    6 DO 7 I=3,N1
      IF(FP(I)-FP(IH).GT.0.0)   GO TO 8
      IF(FP(I)-FP(IL).GE.0.0)   GO TO 7
      IL = I
      GO TO 7
    8 IH = I
    7 CONTINUE
      XN = N
```

```
      50 K1 = NN
         DO 9 I=1,N
         K = I
         S = 0.0
         DO 10 J=1,N1
         IF(J-IH.EQ.0)    GO TO 10
         S = S + P(K)
      10 K = K + N
         K1 = K1 + 1
       9 P(K1) = S / XN
         K = NN + 1
         DO 11 I=1,N
         X(I) = P(K)
      11 K = K + 1
         CALL CALC(N,X,F0,&100)
         WRITE(OUT,12341)   IL, IH, IS
         COMMON /FIVE/ IN, OUT
         INTEGER OUT
   12341 FORMAT('+',100X,3I3,3X,'TEST')
         S = 0.0
         DO 12 I=1,N1
      12 S = S + (FP(I) - F0) ** 2
         S = S / XN
         IF(S-EPS.LE.0.00)   GO TO 100
         IF((IH-1).EQ.0)   GO TO 13
         IS = 1
         GO TO 14
      13 IS = 2
      14 DO 15 I=1,N1
         IF(I-IH.EQ.0)   GO TO 15
         IF(FP(I)-FP(IS).LE.0.0)   GO TO 15
         IS = I
      15 CONTINUE
C**** REFLECTION
         K = (IH - 1) * N + 1
         K0 = NN + 1
         DO 16 I=1,N
         X(I) = 2.0 * P(K0) - P(K)
         K = K + 1
      16 K0 = K0 + 1
         K = K - N
         CALL CALC(N,X,F,&100)
         WRITE(OUT,12342)   IL, IH, IS
   12342 FORMAT('+',100X,3I3,3X,'REFLECTION')
         IF(F-FP(IL).GE.0.0)   GO TO 20
C**** EXPANSION
         K0 = NN + 1
         DO 17 I=1,N
         XS(I) = 2.0 * X(I) - P(K0)
      17 K0 = K0 + 1
         CALL CALC(N,XS,FS,&100)
         WRITE(OUT,12343)   IL, IH, IS
   12343 FORMAT('+',100X,3I3,3X,'EXPANSION')
         IF(FS-FP(IL).GE.0.0)   GO TO 18
         DO 19 I=1,N
         P(K) = XS(I)
      19 K = K + 1
         FP(IH) = FS
         IL = IH
         IH = IS
         GO TO 50
      18 IL = IH
         IH = IS
         FP(IL) = F
      21 DO 22 I=1,N
```

```
         P(K) = X(I)
      22 K= K + 1
         GO TO 50
      20 IF(F-FP(IS).GE.0.0)   GO TO 23
         FP(IH) = F
         IH = IS
         GO TO 21
      23 IF(F-FP(IH).GE.0.0)   GO TO 25
         DO 24 I=1,N
         P(K) = X(I)
      24 K = K + 1
         FP(IH) = F
C****CONTRACTION
         K = K - N
      25 K0 = NN + 1
         DO 26 I=1,N
         XS(I) = 0.5 * (P(K) + P(K0))
         K = K + 1
      26 K0 = K0 + 1
         K = K - N
         CALL CALC(N,XS,FS,&100)
         WRITE(OUT,12344)    IL, IH, IS
12344 FORMAT('+',100X,3I3,3X,'CONTRACTION')
         IF(FS-FP(IH).GE.0.0)   GO TO 40
         DO 27 I=1,N
         P(K) = XS(I)
      27 K = K + 1
         FP(IH) = FS
         IF(FP(1)-FP(2).GT.0.0)    GO TO 28
         IH = 2
         GO TO 29
      28 IH = 1
      29 DO 31 I=3,N1
         IF(FP(I)-FP(IH).LE.0.0)    GO TO 31
         IH = I
      31 CONTINUE
         GO TO 50
      40 FP(1) = FP(IL)
         KKK = MOD(ISH,10)
         ISH = ISH + 1
         IF((IL-1).EQ.0)   GO TO 43
         K = (IL - 1) * N
         DO 41 I=1,N
         K = K + 1
         X(I) = P(K)
         P(K) = P(I)
      41 P(I) = X(I)
         IL = 1
      43 K = N
         DO 42 I=2,N1
         DO 42 J=1,N
         K = K + 1
         P(K) = 0.5 * (P(K) + P(J))
         ISH = ISH + 1
      42 CONTINUE
         GO TO 2
     100 IL = 1
         DO 101 I=2,N1
         IF(FP(I)-FP(IL).GE.0.00)   GO TO 101
         IL = I
     101 CONTINUE
C        WRITE(OUT,2349)   IL, F0, (FP(I), I=1,N1)
C2349 FORMAT('0','NELM EXIT',I5,1P7E15.4,//)
         F = F0
         IF(F.LT.FP(IL))   RETURN
         K = (IL - 1) * N
```

```
            DO 102 I=1,N
            K = K + 1
            X(I) = P(K)
    102     CONTINUE
            F = FP(IL)
            RETURN
            END
            SUBROUTINE CALC(M,X,F,*)
            INTEGER  OUT, REFINE
            REAL LVESSL, LBURET
            COMMON /ONE/   PK(7,3), VZERO(3), HVESSL(3), LVESSL(3), HBURET(3),
           *     LBURET(3), EZERO(3), SLOPE(3), REFINE(15), NDPV, NDPB
            COMMON /TWO/ CI(7),CX(2),TT(2),HX(2),TOLC(2),DT(2),DDT(2,2)
            COMMON /THREE/  V(100), E(100), NP
            COMMON /FOUR/   BETA(7), ARRAY(3), TOL, NCONST, NBETAH, JQR(2,7)
            COMMON /FIVE/   IN, OUT, IFAIL, JFAIL, AL10, KOUNT
            DIMENSION X(M)
C
C
            IF(M.EQ.0)   GO TO 20
            KOUNT = KOUNT + 1
            IF(KOUNT.GE.M*50+80)   GO TO 60
C
C     LOAD THE PARAMETERS BEING REFINED.
C
            I = 0
            J = 7
            DO 15 K=1,NCONST
            J = J + 1
            IF(REFINE(J).LT.1)   GO TO 15
            DO 10 L=1,3
     10     ARRAY(L) = PK(K,L)
            CALL LOAD(&50,F,I,M,X,REFINE(J),ARRAY)
            PK(K,1) = ARRAY(1)
     15     CONTINUE
            IF(REFINE(1).GT.0)     CALL LOAD(&56,F,I,M,X,REFINE(1),VZERO)
            IF(REFINE(2).GT.0)     CALL LOAD(&56,F,I,M,X,REFINE(2),HVESSL)
            IF(REFINE(3).GT.0)     CALL LOAD(&56,F,I,M,X,REFINE(3),LVESSL)
            IF(REFINE(4).GT.0)     CALL LOAD(&56,F,I,M,X,REFINE(4),HBURET)
            IF(REFINE(5).GT.0)     CALL LOAD(&56,F,I,M,X,REFINE(5),LBURET)
            IF(REFINE(6).GT.0)     CALL LOAD(&56,F,I,M,X,REFINE(6),EZERO)
            IF(REFINE(7).GT.0)     CALL LOAD(&56,F,I,M,X,REFINE(7),SLOPE)
C
C     CALCULATION OF THE SUM OF SQUARED RESIDUALS.
C
     20     I = 1
            J = NCONST
            ALFH = 0.000
     21     ALFH = ALFH + PK(J,1) * AL10
            BETA(I) = ALFH
            I = I + 1
            J = J - 1
            IF(J.GT.1)   GO TO 21
            BETA(I) = -PK(1,1) * AL10
C
C     IF(KOUNT.GE.4)   GO TO 60
C     WRITE(OUT,10008)   VZERO(1),HVESSL(1),LVESSL(1),HBURET(1),
C    *  LBURET(1), EZERO(1), SLOPE(1)
C     WRITE(OUT,10008)   BETA
C
            THZERO = (HVESSL(1) + LVESSL(1) * NDPV) * VZERO(1)
            THBC = HBURET(1) + LBURET(1) * NDPB
            TLZERO = LVESSL(1) * VZERO(1)
            ALSLP = AL10 / SLOPE(1)
C
```

Appendix A: MAGEC

```
          F = 0.000
          DO 35 I=1,NP
          VOL = VZERO(1) + V(I)
          TT(1) = (THZERO + THBC * V(I)) / VOL
          TT(2) = (TLZERO + LBURET(1) * V(I)) / VOL
          ALFH = (E(I) - EZERO(1)) * ALSLP
          CX(1) = EXP(ALFH)
          ALFL = 1.0000
          DO 31 K=1,NBETAH
       31 ALFL = ALFL + EXP(BETA(K) + ALFH * K)
          CX(2) = TT(2) / ALFL
          ALFL = ALOG(CX(2))
          DO 32 K=1,NBETAH
       32 CI(K) = EXP(BETA(K)+ALFL+ALFH*K)
          CI(NCONST) = EXP(BETA(NCONST)-ALFH)
          THCALC = CX(1)
          DO 33 J=1,NCONST
       33 THCALC = THCALC + CI(J) * JQR(1,J)
          VCALC = (THZERO - THCALC * VZERO(1)) / (THCALC - THBC)
          RESID = VCALC - V(I)
          IF(KOUNT.GT.0)  GO TO 35
    C     WRITE(OUT,10008)   TT, CX, TOL, ALFL
    C     WRITE(OUT,10008)   CI
          CALL ML(NCONST,2,TOL,2,BETA,CI,CX,TT,HX,TOLC,DT,DDT,JQR)
          IF(IFAIL.GT.0)   GO TO 60
          ECALC = EZERO(1) + ALOG(CX(1)) / ALSLP
          ALFH = -ALOG10(CX(1))
          ALFL = ECALC - E(I)
          THCALC = E(I) + SLOPE(1) * ALFH
          SCALC = (EZERO(1) - E(I)) / ALFH
          WRITE(OUT,10001)   V(I), VCALC, E(I), ECALC, RESID, ALFL, ALFH,
         1 CX(2), THCALC, SCALC
          IF(IFAIL.LT.JFAIL)   WRITE(OUT,10005)
          JFAIL = IFAIL
          TT(1) = (THZERO + THBC*(V(I)+0.004)) / VOL
          CALL ML(NCONST,2,TOL,2,BETA,CI,CX,TT,HX,TOLC,DT,DDT,JQR)
          ALFH = ALFH + ALOG10(CX(1))
          ALFH = ABS(ALFH) * 60.0
          IF(ALFH.GT.0.3.AND.KOUNT.LE.0)   WRITE(OUT,10007)
       35 F = F + RESID ** 2
       50 IF(KOUNT.GT.0)   GO TO 55
          IF(IFAIL.NE.0)   GO TO 58
          RETURN
       55 WRITE(OUT,10008)   F, (X(I), I=1,M)
          IFAIL = 0
          JFAIL = 0
       56 RETURN
       58 WRITE(OUT,10006)
       60 WRITE(OUT,10002)
          IF(IFAIL.NE.0)   WRITE(OUT,10003)
          IF(IFAIL.LT.1)   WRITE(OUT,10004)
          IFAIL = 10
          JFAIL = 0
          RETURN 1
    10001 FORMAT(' ',2F9.3,2F10.2,F12.3,F9.3,F13.3,1PE13.2,0PF16.3,F11.3)
    10002 FORMAT(' ','FAILURE IN SUBROUTINE CALC')
    10003 FORMAT(' ','CAUSED BY NON-CONVERGENCE IN SUBROUTINE ML')
    10004 FORMAT(' ','THE MAXIMUM NUMBER OF ITERATIONS HAS BEEN EXCEEDED.')
    10005 FORMAT('+',118X,'?')
    10006 FORMAT('0','A QUESTION MARK INDICATES A')
    10007 FORMAT('+',120X,'UNBUFFERED')
    10008 FORMAT(' ',1P8E12.3)
          END
          SUBROUTINE LOAD(*,F,I,M,X,IREF,ARRAY)
          DIMENSION  X(M), ARRAY(3)
```

```
      I = I + 1
      ARRAY(1) = X(I)
      IF(IREF.EQ.1)   RETURN
      IF(ARRAY(2).LT.X(I).AND.ARRAY(3).GT.X(I))   RETURN
      XA = (ARRAY(3) + ARRAY(2)) / 2.0
      XA = (X(I) - XA) / (ARRAY(3) - ARRAY(2))
      IF(XA.LT.0.000)   XA = - XA
      F = XA * 1.00E10
      RETURN 1
      END
      SUBROUTINE ML(NK,NMBE,TOL,NC,HLNB,CI,CX,TT,HX,TOLC,DT,DDT,JQR)
      DIMENSION HLNB(NK),CI(NK)
      DIMENSION JQR(NMBE,NK)
      DIMENSION CX(NMBE),TT(NMBE),HX(NMBE)
      DIMENSION TOLC(NC),DT(NC),DDT(NC,NC)
      COMMON /FIVE/   JINP, JOUT, IFAIL, JFAIL, AL10, KOUNT
C         THIS ROUTINE CALCULATES ESTIMATES OF THE FREE CONCENTRATIONS O
C         LIGAND ETC.  USING A NUMBER OF MASS-BALANCE EQUATIONS EQUAL TO
C         NUMBER OF UNKNOWNS (THOSE FOR WHICH THERE IS NO POTENTIAL), TH
C         'NEWTON-RAPHSON' METHOD IS USED, WITH FIRST DERIVATIVES ONLY.
C         ESTIMATES ARE ALSO REQUIRED FOR THIS ROUTINE BUT 1.E-07 WILL
C         SUFFICE IF A MORE ACCURATE VALUE IS NOT AVAILABLE.
      NEMF = NMBE - NC
      IF(NEMF.EQ.0)   GO TO 103
      DO 102 I=1,NEMF
      IPNC = I + NC
  102 HX(IPNC)=ALOG(CX(IPNC))
  103 NCICL = 0
C         A CYCLE COUNTER. 100 CYCLES ARE PERMITTED AS MAXIMIMUM.
      DO 105 I=1,NC
C         TOLC(I) PROVIDES A RELATIVE TOLERANCE FOR USE WITH THE
C         CONVERGENCE CRITERION
  105 TOLC(I)=ABS(TT(I))*TOL
  121 NCICL=NCICL+1
      DO 125 J=1,NC
C         XC(J) IS ONE OF THE UNKNOWN CONCENTRATIONS THAT ARE BEING CALC
C         AS IT CANNOT TAKE A NEGATIVE VALUE, THE STEP LENGTH OF THE COR
C         VECTOR HX IS REDUCED SO THAT NONE OF THEM TAKES A NEGATIVE VAL
  122 IF(CX(J))123,123,125
  123 DO 124 I=1,NC
      HX(I)=0.5*HX(I)
  124 CX(I)=CX(I)-HX(I)
      GO TO 122
  125 CONTINUE
      DO 126 I=1,NC
      HX(I)=ALOG(CX(I))
C         DT(I) IS THE DIFFERENCE BETWEEN T OBSERVED AND T CALCULATED
C         FOR THE MASS-BALANCE EQUATION (I), I.E. IT IS THE RESIDUAL.
  126 DT(I)=CX(I)-TT(I)
C
C     CHANGES AS RECOMMENDED BY LEGGETT   (TALANTA, 1978)
C
      DO 128 J=1,NK
      W=HLNB(J)
      DO 127 I=1,NMBE
  127 W=W+HX(I)*JQR(I,J)
C         CI(J) IS THE CONCENTRATION OF THE SPECIES (J) DEFINED BY
C         THE INDICES IN JQR
      CI(J)=EXP(W)
      DO 128 I=1,NC
  128 DT(I)=DT(I)+JQR(I,J)*CI(J)
      IF(KOUNT.GT.1000)   RETURN
      DO 129 I=1,NC
C         CONVERGENCE CRITERION. WHEN ALL THE MASS-BALANCE EQUATIONS
C         SATISFIED TO THE REQUIRED RELATIVE TOLERANCE, CONTROL IS
```

```
C              PASSED BACK TO THE CALLING PROGRAM.
         IF(ABS(DT(I))-TOLC(I))129,129,131
     129 CONTINUE
         GO TO 190
     131 DO 152 I=1,NC
         DO 151 J=1,NC
C              DDT IS THE JACOBIAN FOR THE SYSTEM, AND IT IS SYMMETRICAL AND
C              SQUARE.  ITS ELEMENTS ARE THE RELATIVE DERIVATIVES, SO THAT
C              THEY ARE OBTAINED DIRECTLY FROM THE CONCENTRATION TERMS
C              PREVIOUSLY CALCULATED.
         DDT(I,J)=0.
         DO 151 L=1,NK
         IF(JQR(I,L))149,151,149
     149 IF(JQR(J,L))150,151,150
     150 W=JQR(I,L)*JQR(J,L)*CI(L)
         DDT(I,J)=DDT(I,J)+W
     151 CONTINUE
     152 DDT(I,I)=DDT(I,I)+CX(I)
         CALL LINEQ(DDT,NC,DT,4)
         IF (IFAIL) 160,160,190
C              DT CONTAINS THE RELATIVE CORRECTIONS TO THE PARAMETERS.
C                    HX WILL CONTAIN THE ABSOLUTE CORRECTIONS.
     160 DO 165 I=1,NC
         HX(I)=-DT(I)*CX(I)
     165 CX(I)=CX(I)+HX(I)
C              IF 100 CYCLES HAVE BEEN EXCEEDED CONTROL IS RETURNED TO CALC.
         IF(NCICL.LT.101)   GO TO 121
         IFAIL = IFAIL - 1
     190 RETURN
         END
         SUBROUTINE LINEQ(A,N,B,KFAIL)
         DIMENSION   A(N,N), B(N)
         COMMON /FIVE/  IN, OUT, IFAIL, JFAIL, AL10, KOUNT
C              SOLVES THE N SIMULTANEOUS LINEAR EQUATIONS A*X=B WITH M RIGHT-
C              SIDES IN B. THE SOLUTION VECTORS ARE LEFT IN B AND THE MATRIX
C              REPLACED BY ITS INVERSE. AFTER CHOLESKI FACTORING OF A TO GIVE
C              THE FORWARD SUBSTITUTIONS L*Y=B AND L*Z=E AND THE BACKWARD SUB
C              LT*X=Y AND LT*AINV=Z ARE PERFORMED
         IF (N-1) 455,5,9
       5 T=A(1,1)
         IF(T.LE.0)   GO TO 455
       6 A(1,1)=1./T
         B(1) = B(1) / T
         RETURN
       9 DO 80 I=1,N
         I1=I-1
         DO 70 J=I,N
         S=A(I,J)
         IF (I1) 10,30,10
      10 DO 20 K=1,I1
      20 S=S-A(I,K)*A(J,K)
      30 X=S
         IF (J-I) 60,40,60
      40 IF (X) 45,45,50
      45 IFAIL=KFAIL
         GO TO 400
      50 A(I,I)=1./SQRT(X)
         GO TO 70
      60 A(J,I)=X*A(I,I)
      70 CONTINUE
      80 CONTINUE
C              FORWARD SUBSTITUTION ON RIGHT HAND SIDES
         B(1) = B(1) * A(1,1)
         DO 120 I=2,N
         I1=I-1
```

```
      S = B(I)
      DO 110 K=1,I1
110   S = S - A(I,K) * B(K)
120   B(I) = S * A(I,I)
C           FORWARD SUBSTITUTION FOR INVERSION
      DO 170 J=1,N
      J1=J+1
      IF (J1-N) 140,140,170
140   DO 160 I=J1,N
      I1=I-1
      S=0.
      DO 150 K=J,I1
150   S=S-A(I,K)*A(J,K)
160   A(J,I)=S*A(I,I)
170   CONTINUE
C           BACKWARD SUBSTITUTION
      B(N) = B(N) * A(N,N)
      DO 220 J=1,N
220   A(J,N)=A(J,N)*A(N,N)
      DO 290 II=2,N
      I=N-II+1
      T=A(I,I)
      I1=I+1
      S = B(I)
      DO 240 K=I1,N
240   S = S - A(K,I) * B(K)
245   B(I) = S * T
      DO 280 J=1,I
      S=A(J,I)
      DO 270 K=I1,N
270   S=S-A(K,I)*A(J,K)
      A(J,I)=S*T
280   CONTINUE
290   CONTINUE
      DO 300 I=2,N
      I1=I-1
      DO 300 J=1,I1
300   A(I,J)=A(J,I)
400   RETURN
C
455   IFAIL=KFAIL
      RETURN
      END
```

C **
C
C
C
C   SUBPROGRAM CALIBT
C   -----------------
C
C
C
C   CALIBRATION AND ANALYSIS
C   OF TITRATION DATA BY
C   LINEAR BEST-FIT TECHNIQUES
C
C
C
C **
C
C
C
C   THIS PROGRAM CALCULATES THE PH (AS THE NEGATIVE LOG OF THE FREE
C   HYDROGEN CONCENTRATION) AT EACH POINT IN A TITRATION AND THEN
C   DETERMINES THE GRADIENT AND INTERCEPT OF THE LEAST SQUARES
C   BEST-FIT WITH RESPECT TO THE GLASS ELECTRODE RESPONSE.
C   THE CALIBRATION IS IMPROVED BY OPTIMISATION OF THE REAGENT

Appendix A: MAGEC

```
C         CONCENTRATIONS OR THE RELEVANT EQUILIBRIUM CONSTANTS.
C         A GRAN PLOT ANALYSIS IS PROVIDED FOR COMPARISON.
C
C
C
C         THE FOLLOWING KINDS OF TITRATION, ALL INVOLVING MONOBASIC
C         REACTANTS, ARE APPLICABLE.
C
C                    (1)    STRONG ACID VERSUS STRONG BASE.
C                    (2)    WEAK ACID VERSUS STRONG BASE.
C                    (3)    STRONG BASE VERSUS STRONG ACID.
C                    (4)    WEAK BASE VERSUS STRONG ACID.
C
C         STRONG ACID OR BASE MAY BE ADDED TO A CORRESPONDING WEAK ACID
C         OR WEAK BASE IN THE VESSEL PRIOR TO TITRATION.  STANDARD ADDITIONS
C         ARE TREATED AS SUBSETS OF THE ABOVE TITRATION SCHEMES.
C
C
C
C         THE NERNST SIGN CONVENTION IS TAKEN TO BE POSITIVE
C         (I.E. INCREASING PH GIVES DECREASING EMF VALUES.)
C         TITRATION VOLUMES MUST INCREASE MONOTONICALLY.
C
C
C
C         WITH STRONG ACID / STRONG BASE TITRATIONS, A SCAN OF PKW CAN BE
C         IMPLEMENTED TO ESTIMATE AN OPTIMUM VALUE.
C         IT IS IMPORTANT TO OBTAIN AGREEMENT BETWEEN:
C         (1) THE OBSERVED AND THEORETICAL SLOPE FOR THE ELECTRODE RESPONSE
C         AND (2) THE VALUES OF EZERO FOUND FOR DATA CORRESPONDING TO THE
C         ACID RANGE COMPARED WITH THAT COVERING ALL BUFFERED POINTS.
C         THESE ARE BETTER CRITERIA THAN A MINIMUM STANDARD DEVIATION.
C         A SCAN OF THE LIGAND PROTONATION CONSTANT IS ALSO POSSIBLE.
C
C***************************************************************
C
C
C
C
      SUBROUTINE CALIBT(VZERO,EZERO,SLOPE,THZERO,TLZERO,AHZERO,
     1    PKWMIN,PKWMAX,BLGMIN,BLGMAX)
C
C
C
      DIMENSION  W(100),H(100),PH(100),PHA(100),PHB(100),G(100),VG(100),
     1    VOLT(100),X(100),Y(100),ET(100),EA(100),EB(100),THT(100)
      INTEGER  IN, OUT, CARD
      REAL NERNST
      LOGICAL   LIGAND, ALKALI, ADJUST
C
C
C
      COMMON /TWO/ CI(7),CX(2),TT(2),HX(2),TOLC(2),DT(2),DDT(2)
      COMMON /THREE/  V(100), E(100), N
      COMMON /FOUR/   BETA(7), ARRAY(3), TOL, NCONST, NBETAH, JQR(2,7)
      COMMON /FIVE/   IN, OUT, IFAIL, JFAIL, AL10, KOUNT
      COMMON /SIX/    CARD(20), P(63)
C
      EQUIVALENCE (P(1),W(1))
C
C
C
  200 FORMAT('1')
  201 FORMAT('0')
```

```
  210 FORMAT(/'0',44X,'SUBPROGRAM CALIBT',/45X,17(1H-),////'0',20A4,//)
  211 FORMAT('0','THE INITIAL VOLUME IS',F7.2,/
     1' ','THE EXPECTED ELECTRODE INTERCEPT IS',F8.2,/
     2' ','THE EXPECTED NERNSTIAN SLOPE IS',F7.2)
  212 FORMAT('0','THE ACID CONCENTRATION IN THE TITRATION VESSEL IS',
     11PE11.3,/' THE LIGAND CONCENTRATION IN THE TITRATION VESSEL IS',
     21PE11.3,/' THE ACID CONCENTRATION IN THE BURETTE IS',1PE11.3)
  213 FORMAT(' ','THE END POINT IS EXPECTED AT',F7.3)
  215 FORMAT(' ','THE VALUE OF PKW IS',F7.3,/' ',26(1H-),/)
  216 FORMAT(' ','THE VALUE OF LOG BETAH IS',F7.3,/' ',31(1H-),/)
  220 FORMAT('0','TITRATION POINT NUMBER',I3,' IS IN ERROR',15X,2F12.2)
  222 FORMAT('0','INITIAL CONCENTRATION DATA PROBABLY INCORRECT',//)
  231 FORMAT('0','THE GRAN-PLOT VALUES AT EACH POINT ARE:',11X,'I',7X,
     1  'VOL',6X,'EMF',8X,'V(I)',12X,'G(I)',///)
  235 FORMAT('0','PKW IS IN ERROR')
  236 FORMAT('0','THE VALUE OF LOG BETAH IS IN ERROR')
  240 FORMAT('0','THE SEARCH FOR UNSUITABLE POINTS HAS REMOVED ')
  241 FORMAT(' ',45X,I6,2F10.2,7(1PE15.4))
  245 FORMAT('+',45X,'NONE')
  250 FORMAT(' (50% OF THE POINTS, TAKEN ABOUT THE MIDDLE OF THE SET)')
  251 FORMAT(' ','BEST-FIT INVOLVING',I3,' POINTS GIVES EZERO =',
     1  F7.1,' (',F4.2,') AND A SLOPE =',F7.2,' (',F4.2,')',/
     2  ' ','OVERALL STANDARD DEVIATION =',1PE11.3)
  252 FORMAT(///'0','USING DATA BEFORE THE ENDPOINT,')
  253 FORMAT('0','USING DATA AFTER THE ENDPOINT,')
  254 FORMAT('0','USING ALL THE BUFFERED DATA,')
  255 FORMAT(///'0',8X,'VOLUME'9X,'TH',12X,'PH',10X,'E(OBS.)',
     1 7X,'E(CALC.)',8X,'RESIDUAL',5X,'PKW(CALC.)',4X,'EZERO(CALC.)',//)
  256 FORMAT(' ',I3,F10.2,1PE15.3,0PF12.3,5F15.2)
  257 FORMAT('0','PKW(CALC.) IS OBTAINED USING THE CURRENT VALUE FOR',
     1  ' EZERO (',F6.1,') AND THE NERNSTIAN SLOPE.  AVERAGE =',F7.3)
  258 FORMAT('0','EZERO(CALC.) IS OBTAINED USING THE CURRENT VALUE FOR',
     1  ' PKW (',F6.2,') AND THE NERNSTIAN SLOPE.  AVERAGE =',F7.1,///)
  260 FORMAT('1',/'0',10X,35(1H*),5X,'NEW SCAN ITERATION',5X,35(1H*),/)
  261 FORMAT('+','NOT REQUIRED.',/
     1  'THE RESULTS ARE INDEPENDENT OF THE VALUE OF PKW.')
  262 FORMAT(//////'0',44X,'CALIBT CONCENTRATION ADJUSTMENT')
  263 FORMAT(' ',43X,'AND TO DETERMINE A VALUE FOR PKW.')
  264 FORMAT(///'0','WITH A VESSEL-ACID CONCENTRATION OF',1PE11.3,/' ',
     1  'AND A BURETTE-ACID CONCENTRATION OF',1PE11.3,/' ',
     2  'THE FOLLOWING RESULTS ARE OBTAINED',////)
  265 FORMAT(' ','THE ENDPOINT NOW OCCURS AT',F7.3,///)
  266 FORMAT(' ',36X,'TO DETERMINE THE BURETTE-ACID CONCENTRATION.',//)
  267 FORMAT(' ',37X,'TO DETERMINE THE VESSEL-ACID CONCENTRATION')
C
C
C
   10 WRITE(OUT,200)
      LIGAND = .FALSE.
      ALKALI = .FALSE.
      ADJUST = .FALSE.
      NADJ = -10
C
      WRITE(OUT,210)    CARD
      NERNST = SLOPE
      WRITE(OUT,211)    VZERO, EZERO, NERNST
      IF(AHZERO.GT.0.000)   ALKALI = .TRUE.
      DUMMY = TLZERO
      IF(ALKALI)   DUMMY = -DUMMY
      ENDPT = -VZERO * (THZERO + DUMMY) / AHZERO
      ARESET = AHZERO
      TRESET = THZERO
      ERESET = ENDPT
      WRITE(OUT,212)    THZERO, TLZERO, AHZERO
      WRITE(OUT,213)    ENDPT
      IF(TLZERO.GT..1E-7)   LIGAND = .TRUE.
```

Appendix A: MAGEC

```
      PKW = PKWMIN
      WRITE(OUT,215)   PKW
      PKWINC = (PKWMAX - PKWMIN) / 5.000
      IF(PKWINC.LT.0.000)   PKWINC = 0.000
      BLGINC = 0.000
      IF(.NOT.LIGAND)   GO TO 16
      BLG = BLGMIN
      WRITE(OUT,216)   BLG
      IF(BLGMAX.GT.BLGMIN)   BLGINC = (BLGMAX - BLGMIN) / 5.000
      IF(BLGINC.GT.0.000)   PKWINC = 0.000
   16 CONTINUE
C
C
C
      DO 20 I=2,N
      IF(V(I).LT.V(I-1))   GO TO 21
   20 CONTINUE
      GO TO 30
   21 WRITE(OUT,220) I, V(I), E(I)
      IF(I.EQ.2)   WRITE(OUT,222)
      GO TO 65
C
C
C
   30 DO 31 I=1,N
   31 H(I) = 10.000 ** ((E(I) - EZERO) / SLOPE)
      WRITE(OUT,231)
      IG = 1
      DO 32 I=1,N
      IF(LIGAND.AND.V(I).LT.0.00099)   GO TO 32
      IF(V(I).GT.ENDPT)   GO TO 33
      G(IG) = VZERO + V(I)
      IF(LIGAND)   G(IG) = V(I)
      HOLD = H(I)
      IF(ALKALI)   HOLD = 1.000 / HOLD
      G(IG) = G(IG) * HOLD
      VG(IG) = V(I)
      WRITE(OUT,241)   I, V(I), E(I), VG(IG), G(IG)
      IG = IG + 1
   32 CONTINUE
   33 IG = IG - 1
      INIT = 1
      IF(IG.GT.5)   CALL STRAIT(VG,G,W,IG,ENDPT,INIT,IFIN,X,Y,OUT)
      NFLAG = IG + 1
      IF(NFLAG.GT.N-3)   GO TO 35
      IG = N - NFLAG + 2
      IF(NFLAG.GT.1)   WRITE(OUT,231)
      DO 34 I=NFLAG,N
      IG = IG - 1
      G(IG) = VZERO + V(I)
      HOLD = H(I)
      IF(.NOT.ALKALI)   HOLD = 1.000 / HOLD
      G(IG) = G(IG) * HOLD
      VG(IG) = V(I)
   34 WRITE(OUT,241)   I, V(I), E(I), VG(IG), G(IG)
      IG = N - NFLAG + 1
      INIT = N
      IF(IG.GT.5)   CALL STRAIT(VG,G,W,IG,ENDPT,INIT,IFIN,X,Y,OUT)
C
   35 IF(PKW.GT.12.0.AND.PKW.LT.15.00)   GO TO 36
      IF(PKW.GT.0.5.OR.PKW.LT.-0.5)   WRITE(OUT,235)
      GO TO 65
   36 IF(.NOT.LIGAND)   GO TO 37
      IF(BLG.GT.2.00.AND.BLG.LT.12.00)   GO TO 40
      IF(BLG.GT.0.5.OR.BLG.LT.-0.5)   WRITE(OUT,236)
      GO TO 65
```

```
   37 TH2 = THZERO * 0.1
      IF(NADJ.GT.15)   TH2 = TH2 / 5.00
      TH1 = THZERO + TH2
      TH2 = THZERO - TH2
C
C
C
   40 IT = 0
      IA = 0
      IB = 0
      NFLAG = 0
      KWSCAN = 1
      THMOL = THZERO * VZERO
      IF(.NOT.(LIGAND.OR.ADJUST))   WRITE(OUT,215)   PKW
      IF(LIGAND)   WRITE(OUT,216)   BLG
      IF(.NOT.ADJUST)   WRITE(OUT,240)
      IF(.NOT.LIGAND)   GO TO 45
C
      TLMOL = TLZERO * VZERO
      IF(.NOT.ALKALI)   THMOL = THMOL + TLMOL
      DUMMY = -PKW
      WK = 10.00 ** DUMMY
      BETAH = 10.000 ** BLG
      DO 44 I=1,N
      W(I) = 1.000
      VOL = VZERO + V(I)
      TL = TLMOL / VOL
      TH = (THMOL + AHZERO * V(I)) / VOL
      HLO = -1.000
      HHI = -1.000
      HFREE = -1.000
      IF(TH.LE.0.000)   GO TO 41
C
C     ACID APPROXIMATION
C
      HOLD = 1.0 + BETAH * (TL - TH)
      HLO = HOLD ** 2 + 4.0 * BETAH * TH
      HLO = (SQRT(HLO) - HOLD) / (2.0 * BETAH)
      HFREE = HLO
   41 IF(TL.LE.TH)   GO TO 42
C
C     ALKALI APPROXIMATION
C
      HOLD = -(TH + WK * BETAH)
      HHI = HOLD ** 2 + 4.0 * BETAH * (TL - TH) * WK
      HHI = (SQRT(HHI) - HOLD) / (2.0 * BETAH * (TL - TH))
      IF(HFREE.LE.0.000)   HFREE = HHI
      IF(HLO.LT.1.00E-12.OR.HHI.LT.1.00E-12)   GO TO 42
C
      DUMMY = SQRT(WK)
      HFREE = -1.000
      IF(HLO.GT.DUMMY*100.0)   HFREE = HLO
      IF(HHI.LT.DUMMY/100.0)   HFREE = HHI
      IF(HFREE.GT.0.000)   GO TO 42
      HOLD = HLO * HHI / WK
      DUMMY = 1.000 / HOLD
      IF(HOLD.LT.1.000)   HOLD = 1.000
      IF(DUMMY.LT.1.000)   DUMMY = 1.000
      HLO = ALOG(HLO) * HOLD
      HHI = ALOG(HHI) * DUMMY
      HFREE = (HLO + HHI) / (HOLD + DUMMY)
      HFREE = EXP(HFREE)
   42 IF(HFREE.LT.1.0E-12)   GO TO 43
      FREEL = TL / (1.0 + BETAH * HFREE)
      FREELH = TL - FREEL
```

```
          IF((FREEL.LT.1.0E-3.OR.FREELH.LT.1.0E-3).AND.
     1    (HFREE.GT.1.0E-11.AND.HFREE.LT.1.0E-3))     GO TO 43
          IT = IT + 1
          PH(IT) = -ALOG10(HFREE)
          VOLT(IT) = V(I)
          THT(IT) = TH
          ET(IT) = E(I)
          IF(V(I).GT.ENDPT)   GO TO 421
          IA = IA + 1
          EA(IA) = E(I)
          PHA(IA) = PH(IT)
          GO TO 44
   421    IB = IB + 1
          EB(IB) = E(I)
          PHB(IB) = PH(IT)
          GO TO 44
    43    IT = IT + 1
          IF(IT.EQ.I)   WRITE(OUT,201)
          IT = IT - 1
          WRITE(OUT,241)   I, V(I), E(I)
    44    CONTINUE
          IF(IT.EQ.N)   WRITE(OUT,245)
          GO TO 52
C
    45    DO 48 I=1,N
          W(I) = 1.0000
          VOL = VZERO + V(I)
          TH = (THMOL + AHZERO * V(I)) / VOL
          IF(TH.LT.9.999E-4)   GO TO 46
          KWSCAN = 1
          IT = IT + 1
          IA = IA + 1
          ET(IT) = E(I)
          EA(IA) = E(I)
          PH(IT) = -ALOG10(TH)
          PHA(IA) = PH(IT)
          VOLT(IT) = V(I)
          THT(IT) = TH
          GO TO 48
    46    IF(TH.LE.-9.999E-4.AND.TH.GT.-0.01)   GO TO 47
          IF(ADJUST)   GO TO 48
          IT = IT + 1
          IF(IT.EQ.I)   WRITE(OUT,201)
          IT = IT - 1
          WRITE(OUT,241)   I, V(I), E(I)
          GO TO 48
    47    IT = IT + 1
          IB = IB + 1
          ET(IT) = E(I)
          EB(IB) = E(I)
          PH(IT) = PKW + ALOG10(-TH)
          PHB(IB) = PH(IT)
          VOLT(IT) = V(I)
          THT(IT) = TH
    48    CONTINUE
C
C
C
    50    IF(.NOT.ADJUST.AND.IT.EQ.N)   WRITE(OUT,245)
          IF(IT.LE.0)   STOP 50
          IF(IT.LE.5)   GO TO 54
          IF(IA.LT.5.OR.IB.LT.5)   GO TO 52
          CALL LINFIT(PHA,EA,W,IA,SLOPE,EZERO,SXSLP,SXINT,STDDEV)
          SLOPE = -SLOPE
          ACIDSL = SLOPE
```

```
              IF(ADJUST)   GO TO 52
              IF(.NOT.ALKALI.OR.LIGAND)   WRITE(OUT,252)
              IF(.NOT.LIGAND.AND.ALKALI)   WRITE(OUT,253)
              IF(IB.EQ.0)   WRITE(OUT,250)
              WRITE(OUT,251)     IA, EZERO, SXINT, SLOPE, SXSLP, STDDEV
C
              CALL LINFIT(PHB,EB,W,IB,SLOPE,EZERO,SXSLP,SXINT,STDDEV)
              SLOPE = -SLOPE
              IF(.NOT.ALKALI.OR.LIGAND)   WRITE(OUT,253)
              IF(.NOT.LIGAND.AND.ALKALI)   WRITE(OUT,252)
              IF(IA.EQ.0)   WRITE(OUT,250)
              WRITE(OUT,251)     IB, EZERO, SXINT, SLOPE, SXSLP, STDDEV
C
           52 IF(ADJUST.AND.NADJ.LT.10.AND.IA.GE.5.AND.IB.GE.5)   GO TO 59
              CALL LINFIT(PH,ET,W,IT,SLOPE,EZERO,SXSLP,SXINT,STDDEV)
              SLOPE = -SLOPE
              IF(IA.LT.5.OR.IB.LT.5)   ACIDSL = SLOPE
              BASESL = SLOPE
              IF(ADJUST)   GO TO 59
              WRITE(OUT,254)
              WRITE(OUT,251)     IT, EZERO, SXINT, SLOPE, SXSLP, STDDEV
C
C
C
           54 WRITE(OUT,255)
              HOLD = 0.000
              DUMMY = 0.000
              PKWCLC = 0.000
              IF(IA.EQ.0)   IA = IT + 1
              DO 55 I=1,IT
              X(I) = PH(I)
              Y(I) = ET(I)
              ECALC = EZERO - SLOPE * PH(I)
              ERESID = ET(I) - ECALC
              EZOCLC = ET(I) + NERNST * PH(I)
              HOLD = HOLD + EZOCLC
              IF(I.LE.IA)   GO TO 55
              IF(I.EQ.IA+1)   WRITE(OUT,201)
              PKWCLC = ((EZERO - ET(I)) / NERNST) - PH(I) + PKW
              DUMMY = DUMMY + (1.0 / 10.00 ** PKWCLC)
           55 WRITE(OUT,256)   I, VOLT(I), THT(I), PH(I), ET(I), ECALC, ERESID,
             1    PKWCLC, EZOCLC
C
              WRITE(OUT,201)
              IF(IT.LE.5.OR.IA.LT.5)   GO TO 65
              IF(IB.LT.3)   GO TO 56
              PKWCLC = DUMMY / FLOAT(IB)
              PKWCLC = -ALOG10(PKWCLC)
              WRITE(OUT,257)   EZERO, PKWCLC
           56 EZOCLC = HOLD / FLOAT(IT)
              WRITE(OUT,258)   PKW, EZOCLC
C
C
C
           59 IF(LIGAND)   GO TO 60
              NADJ = NADJ + 1
              IF(NADJ-11)   591, 592, 594
          591 ADJUST = .TRUE.
              ACIDSL = ACIDSL - NERNST
              IF(NERNST.LT.0.00)   ACIDSL = -ACIDSL
              IF(ACIDSL.GT.0.000)   TH1 = THZERO
              IF(ACIDSL.LT.0.000)   TH2 = THZERO
              THZERO = (TH1 + TH2) / 2.000
              GO TO 40
          592 IF(.NOT.ADJUST)   GO TO 593
```

Appendix A: MAGEC

```
          ENDPT = -VZERO * THZERO / AHZERO
          WRITE(OUT,262)
          WRITE(OUT,267)
          IF(IB.GT.3)   WRITE(OUT,263)
          WRITE(OUT,201)
          WRITE(OUT,264)   THZERO, AHZERO
          WRITE(OUT,265)   ENDPT
          ADJUST = .FALSE.
          NADJ = 10
          GO TO 40
      593 ADJUST = .TRUE.
          TH2 = AHZERO * 0.1
          IF(NADJ.GT.15)   TH2 = TH2 / 5.00
          TH1 = AHZERO + TH2
          TH2 = AHZERO - TH2
          NADJ = NADJ + 1
      594 IF(NADJ-21)    597, 37, 595
      595 IF(NADJ-31)    591, 593, 596
      596 IF(NADJ-40)    597, 598, 60
      597 BASESL = BASESL - NERNST
          IF(NERNST.LT.0.00)   BASESL = -BASESL
          IF(BASESL.GT.0.000)   TH2 = AHZERO
          IF(BASESL.LT.0.000)   TH1 = AHZERO
          AHZERO = (TH1 + TH2) / 2.000
          THZERO = -AHZERO * ENDPT / VZERO
          GO TO 40
      598 IF(IA.LT.5.OR.IB.LT.5)   GO TO 65
          WRITE(OUT,262)
          WRITE(OUT,266)
          WRITE(OUT,264)   THZERO, AHZERO
          WRITE(OUT,265)   ENDPT
          ADJUST = .FALSE.
          GO TO 40
C
C
C
       60 IF(PKWINC.LE.0.0001)   GO TO 62
          NADJ = 0
          PKW = PKW + PKWINC
          IF(PKW.GT.PKWMAX)   GO TO 62
          WRITE(OUT,260)
          IF(KWSCAN.EQ.0)   WRITE(OUT,260)
          IF(KWSCAN.EQ.0)   GO TO 62
          THZERO = TRESET
          AHZERO = ARESET
          ENDPT = ERESET
          WRITE(OUT,264)   THZERO, AHZERO
          WRITE(OUT,213)   ENDPT
          GO TO 35
C
       62 IF(.NOT.LIGAND.OR.BLGINC.LE.0.0001)   GO TO 65
          BLG = BLG + BLGINC
          IF(BLG.GT.BLGMAX)   GO TO 65
          WRITE(OUT,260)
          GO TO 35
C
       65 WRITE(OUT,200)
          RETURN
          END
C
C
C
C
C
```

```
C       THE FOLLOWING SUBROUTINE SEARCHES DATA STORED IN THE ARRAYS
C       NAMED V AND G TO FIND THE SEGMENT GIVING THE STRAIGHTEST LINE.
C       ONLY THOSE LINES WITH ABSSICA INTERCEPTS WITHIN PTOL PERCENT OF
C       THE ESTIMATED END POINT VALUE ARE ACCEPTED;  IF THIS CONDITION IS
C       SATISFIED, THE SEGMENT GIVING THE LOWEST STANDARD DEVIATION IS
C       LOCATED;  OTHERWISE, THE LINE FOUND IS THAT GIVING THE CLOSEST
C       AGREEMENT WITH THE END POINT ESTIMATE.   SUBROUTINE LINFIT IS
C       CALLED BETWEEN 20 AND 50 TIMES;  ITER DETERMINES HOW FINELY
C       THE DATA IS DIVIDED INTO LINE SEGMENTS.
C
C
C
C
C
C
        SUBROUTINE STRAIT(V,G,W,N,ENDPT,INIT,IFIN,X,Y,OUT)
        DIMENSION  V(N),  G(N),  W(N),  X(N),  Y(N)
        INTEGER OUT
C
C
C
   201  FORMAT(///'0','GRAN-PLOT EXTRAPOLATIONS',/' ',24(1H*),///'0',
       1  2X,'SEGMENT',8X,'END PT.',5X,'STD. DEV.',//)
   202  FORMAT(' ',I3,2X,'-',I4,4X,F10.4,3(1PE15.4))
   203  FORMAT(//'0','THE STRAIGHTEST SEGMENT GIVES',3X,F10.4,/'0',
       1  'THE MINIMUM STD. DEVIATION GIVES',3X,F10.4)
   205  FORMAT('+',48X,'OMITTED BECAUSE THE ESTIMATED ENDPOINT IS OUTSIDE
       1THE PERMITTED RANGE OF', F5.1,'%')
   207  FORMAT('0','THE WEIGHTED AVERAGE GIVES',3X,F8.4,/'0',
       1  'THE ACTUAL AVERAGE GIVES',3X,F8.4,////)
   210  FORMAT(' FAILED TO FIND THE ENDPOINT',//)
   211  FORMAT(' CLOSEST AGREEMENT WITH THE ESTIMATED VALUE IS WITHIN',
       1  F6.1,'%',//)
   215  FORMAT('0ANOTHER ATTEMPT WITH INCREASED ENDPOINT RANGE',///)
   218  FORMAT('0',I5,' POINTS HAVE BEEN OMITTED BECAUSE THEIR ESTIMATED
       1ENDPOINTS ARE OUTSIDE A RANGE OF',F6.1,'%',///)
C
C
C
        PTOL = 5.0
        ISTART = INIT
     5  ITER = 5
        IF(ISTART.EQ.1)  ISTART = -1
        INIT = 0
        ENDOLD = 1.000E19
        STDOLD = 1.000E19
        STDSTD = 1.000E19
        SAVINT = 0.000
        SAVSLP = 0.000
        KOUNT = 0
        WGTAVE = 0.000
        ENDAVE = 0.000
        STDAVE = 0.000
        NDUD = 0
C
        IF(N.LT.6)   STOP
        MMIN = 5
        IF(N.GT.12)  MMIN = 6
        IF(N.GT.25)  MMIN = 7
        IF(N.GT.50)  MMIN = 9
        NM1 = N - 1
        MINC = (N - MMIN) / ITER
        IF(MINC.LT.1)  MINC = 1
        DO 10 I=1,N
    10  W(I) = 1.000
        IF(PTOL.LT.7.00)  WRITE(OUT,201)
```

Appendix A: MAGEC

```
          DO 35 M=MMIN,NM1,MINC
          MP1 = M + 1
          IINC = ((N - M) / ITER)
          IF(IINC.LT.1)   IINC = 1
C
          DO 35 I=M,N,IINC
          DO 20 J=1,M
          K = I + J - M
          X(J) = V(K)
       20 Y(J) = G(K)
          J - I - M + 1
          CALL LINFIT(X,Y,W,M,SLOPE,XINT,SXSLP,SXINT,STDDEV)
          ENDTRY = -XINT / SLOPE
          ENDAVE = ENDAVE + ENDTRY
          WGTAVE = WGTAVE + ENDTRY / STDDEV
          STDAVE = STDAVE + 1.000 / STDDEV
          KOUNT = KOUNT + 1
          K = IABS(ISTART-J+1)
          L = IABS(ISTART-I+1)
          IF(PTOL.LT.7.00)   WRITE(OUT,202)   K,  L,  ENDTRY, STDDEV
          ENDNEW = (ABS(ENDTRY-ENDPT) / ENDPT) * 100.0
          IF(ENDNEW.LT.PTOL)   GO TO 30
          IF(PTOL.LT.7.00)   WRITE(OUT,205)    PTOL
          ENDAVE = ENDAVE - ENDTRY
          KOUNT = KOUNT - 1
          WGTAVE = WGTAVE - ENDTRY / STDDEV
          STDAVE = STDAVE - 1.000 / STDDEV
          NDUD = NDUD + 1
          IF(ENDNEW.GT.ENDOLD)   GO TO 35
          ENDOLD = ENDNEW
          GO TO 32
       30 ENDOLD = 0.0000
          IF(STDDEV/FLOAT(I-J+2).GT.STDOLD)    GO TO 34
       32 INIT = J
          IFIN = I
          SAVINT = XINT
          SAVSLP = SLOPE
          STDOLD = STDDEV / FLOAT(I-J+2)
       34 IF(STDDEV.GT.STDSTD)   GO TO 35
          STDSTD = STDDEV
          SAVEND = ENDTRY
       35 CONTINUE
          IF(KOUNT.GE.1)   GO TO 38
          WRITE(OUT,210)
          IF(ENDOLD.LT.1.0E2)   WRITE(OUT,211)   ENDOLD
       36 IF(PTOL.GT.7.0)    GO TO 39
          PTOL = PTOL * 2
          IF(ENDOLD.GT.PTOL) RETURN
          WRITE(OUT,215)
          GO TO 5
       38 STDDEV = STDOLD * FLOAT(IFIN-INIT+2)
          ENDTRY = -SAVINT / SAVSLP
          ENDAVE = ENDAVE / FLOAT(KOUNT)
          WGTAVE = WGTAVE / STDAVE
          WRITE(OUT,203)   ENDTRY, SAVEND
          WRITE(OUT,207)   WGTAVE,   ENDAVE
          IF(FLOAT(NDUD).GT.FLOAT(KOUNT)/2.0)    GO TO 36
          RETURN
       39 WRITE(OUT,218)   NDUD, PTOL
          RETURN
          END
C
C
```

```
C     LINEAR LEAST SQUARES FIT ROUTINE
C     --------------------------------
C
C
C     DEFINITION OF VARIABLE NAMES:
C
C
C        X, Y           = DATA ARRAYS
C        W              = ARRAY FOR WEIGHTING FACTORS
C        N              = NUMBER OF DATA POINTS
C        NW             = FLAG FOR WEIGHTING
C        XINT, XSLOPE   = INTERCEPT AND SLOPE OF LEAST SQUARES LINE
C        SXINT, SXSLP   = STANDARD DEVIATIONS OF XINT AND XSLOPE
C        STDDEV         = STANDARD DEVIATION OF THE LINE FIT
C
C
C
      SUBROUTINE LINFIT(X,Y,W,N,XSLOPE,XINT,SXSLP,SXINT,STDDEV)
      DIMENSION X(N), Y(N), W(N)
      WW = 0.000
      WX = 0.000
      WY = 0.000
      WXY = 0.000
      WXX = 0.000
      WYY = 0.000
      DO 10 I=1,N
      WW = WW + W(I)
      WY = WY + W(I) * Y(I)
      WX = WX + W(I) * X(I)
      WXY = WXY + W(I) * X(I) * Y(I)
      WXX = WXX + W(I) * X(I) ** 2
   10 WYY = WYY + W(I) * Y(I) ** 2
      DENOM = WW * WXX - WX ** 2
      XSLOPE = (WW * WXY - WX * WY) / DENOM
      XINT = (WXX * WY - WX * WXY) / DENOM
      VSUM = 0.000
      DO  20  I=1,N
   20 VSUM = VSUM + (Y(I) - XINT - XSLOPE * X(I)) ** 2
      SS = VSUM / FLOAT(N-2)
      STDDEV = SQRT(SS)
      SXINT = SQRT((SS/WW)*(1.0+((WX**2)/DENOM)))
      SXSLP = SQRT(SS*WW/DENOM)
      RETURN
      END
```

Appendix B. The ACBA program (Arena et al. 1979)

Data input

The following data input is required:
(1) 1 card Format (20A4): TITLE; descriptive title (job)
(2) 1 card Format (3I2.12A5): NC, NL, NVAR, (PVAR (I), I = 1, NVAR)
NC: maximum number of iteration cycles to be performed
NL: number of ligands
NVAR: number of common parameters to be refined
PVAR (I): symbols of common parameters to be refined
COH: initial concentration of the strong acid (positive) or base (negative)
CTH: concentration of the strong acid (positive) or base (negative) in the titrant
EO: standard potential
KWL: $\lg K_W = -pK_W$
JA: j_a (coefficient of junction potential $E_j = j_a h$)
SL: s_L
COL 1, COL 2 $\cdots$ initial concentration of ligands
CTL 1, CTL 2 $\cdots$ concentrations of the ligands in the titrant
K11, K21 $\cdots$ K31 $\cdots$: $\lg K_{ij}^H$ (stepwise association constants)
(3) NL cards FORMAT (5F10.0): ((KL(I, K), K = 1,5), I = I,NL: (only if NL > 0)

$$KL(K, K): \lg K_{ij}^H$$

(4) 1 card Format (2F10.0) TEMP. KWL: temperature °C, $-pK_W$.
The following cards for each titration curve:
(5) 1 card Format (20A4): descriptive title (titration curve)
(6) 1 card Format (2F10.0, 3F7.0, F10.0, 2F7.0, I3): VX(1), VX(2), VO VX(3), VX(5), FDILH, VIN, DV, NA:
VX(1) = COH (see item (2))
VX(2) = CTH
VO: initial volume
VX(3) = EO
VX(5) = JA
FDILH: dilution factor for COH (to allow overall refinement of the original concentration of a solution which has been titrated at different dilutions)
VIN = V_0; DV = v_a; NA = N (see item (9))
(7) NL cards Format (2F10, 2I2, F10.0): VX(J), VX(J+1), NNPO(L, K.), NNPT(L, K). FDILLM(L, K): only if NL > 0
VX(J) = COL (see item (2))
VX(J+1) = CTL
NNPO: number of protons in the ligand (titrand)
NNPT: number of protons in the ligand (titrant)
FDILLM: dilution factor for COL (see item (6))
(8) 1 card Format (I2.12A5): NVART, (PVAR(K), K = 1, NVART):
NVART: number of titration parameters to be refined
PVAR(K): symbols of parameters to be refined (see item (2))
(9) cards Format (5(2F7.0, I1): (VV(K)), EE(K), (INDEX(K), K = 1,5); (if DV = 0)
VV(K): titrant volume added in cm^3
EE(K): emf (mV): if EO = 0 EE(K) = pH
INDEXZ(K): = 0 normal; = 1 indicates the end of a titration curve GOTO (5)

(10) cards Format $(-)$:(KE(K), K = 1, NA):without format but expressed as an integer (if DV>0)
KE(K):emf = 10^{-1} KE or (if EO = 0) pH = 10^{-3} KE
The titrant volume added is: $v = V_0 + k \cdot va$ $k = 0 \cdots$ NA (see item (6)) for a titration curve with constant increment of titrant
When K = NA GOTO (5)

(11) 2 blank cards: end of all titration curves
4 blank cards: for the termination of the run
Concentrations: mole/l
Volumes: cm^3
emf: mV

Program listing

```
                                   PROGRAM

       PROGRAM ACBA(INPUT,OUTPUT,TAPE5=INPUT,TAPE6=OUTPUT)           1
C                                                                    2
C  THIS PROGRAM REFINES THE PARAMETERS OF ONE OR MORE ACID-BASE TITRATIONS  3
C                                                                    4
C  CARMELO RIGAMO - SEMINARIO MATEMATICO                             5
C  GIUSEPPE ARENA, ENRICO RIZZARELLI, SILVIO SAMMARTANO - ISTITUTO DI  6
C  CHIMICA GENERALE         UNIVERSITATA DI CATANIA (1977)           7
C                                                                    8
       DIMENSION KVAR(12),B(4,5),NB(4),NNPO(20,4),NNPT(20,4),FDILLM(20,4)  9
      &,FDILHM(20),IVT(20,6),X(20,14),VVO(20),NVT(20),SVQ(20),V(700),E(70  10
      &0),LAST(20),NP(20),IVAR(12),XM(12),BM(4,5),TOLL(12),COL(4),CTL(4),  11
      &FDILL(4),A1M(4),A2M(4),AM(4),DAH(4),PH(700),DV(700),Z(700,4),A2K(4  12
      &,5),A1K(4,5),SH(12),SIGX(12),XPT(34),SIGT(34),XPP(20,14),SIGP(20,1  13
      &4),DEVS(20),RL(20),NPO(4),NPT(4),D(700,12),DER(12)             14
       REAL KP(4,5),LIM(12),NUM,JA,KW                                 15
C                                                                    16
C  CAN VARY COMMON PARAMETERS WITHOUT VARYING THE TITRATION PARAMETERS  17
C  THE JOB STARTS OFF BY VARYING THE COMMON PARAMETERS                18
C                                                                    19
   70 CALL DATI(NC,NL,NVAR,SVQ,KVAR,KP,B,NB,X,VVO,FDILHM,FDILLM,NNPO,  20
      &NNPT,NVT,IVT,LEV2,V,E,LAST,NP,NTP,NTIT,AL10,JB,SVQT)            21
C                                                                    22
C  LEV=1 REFINES THE PARAMETERS TO BE VARIED SIMULTANEOUSLY           23
C  IN ALL TITRATIONS                                                  24
C  LEV=2 REFINES THE PARAMETERS TO BE VARIED IN EACH TITRATION        25
C                                                                    26
       ISDV=0                                                         27
       NCGP=0                                                         28
       LEV=1                                                          29
       NCG=1                                                          30
       NCICP=NC*2                                                     31
   82  NCC=0                                                          32
   81  MLEV=0                                                         33
   54  L=1                                                            34
       K1=1                                                           35
       IF(LEV.EQ.2) GOTO 35                                           36
       IF(ISDV.EQ.1) GOTO 52                                          37
       N=NTP                                                          38
       NV=NVAR                                                        39
       SV2=SVQT                                                       40
       DO 36 K=1,NV                                                   41
   36  IVAR(K)=KVAR(K)                                                42
       GOTO 51                                                        43
   35  IF(NVT(L).EQ.0) GOTO 83                                        44
       NCC=0                                                          45
       N=NP(L)                                                        46
       NV=NVT(L)                                                      47
       SV2=SVQ(L)                                                     48
       DO 37 K=1,NV                                                   49
   37  IVAR(K)=IVT(L,K)                                               50
   51  JAV=0                                                          51
       JEO=0                                                          52
       JVH=0                                                          53
       JSL=0                                                          54
C  MEMORIZES THE ESTIMATES OF PARAMETERS TO BE REFINED                55
```

Appendix B: ACBA

```
      DO 4 I=1,NV                                                  56
      K=IVAR(I)                                                    57
      IF(NCC.GT.0) GOTO 10                                         58
      IF(K.LT.15) GOTO 92                                          59
      L1=(K-10)/5                                                  60
      L2=K-9-L1*5                                                  61
      XM(I)=KP(L1,L2)                                              62
      BM(L1,L2)=B(L1,L2)                                           63
      GOTO 4                                                       64
   92 XM(I)=X(L,K)                                                 65
   10 IF(K.NE.3) GOTO 20                                           66
      JEO=1                                                        67
      GOTO 25                                                      68
   20 IF(K.NE.5) GOTO 71                                           69
      JAV=1                                                        70
      GOTO 25                                                      71
   71 IF(K.NE.6) GOTO 4                                            72
      JSL=1                                                        73
   25 JVH=1                                                        74
    4 CONTINUE                                                     75
C FIXES LIMITS AND TOLERANCES OF THE SHIFTS                        76
   86 DO 73 I=1,NV                                                 77
      K=IVAR(I)                                                    78
      IF(K.LT.15) GOTO 79                                          79
      L1=(K-10)/5                                                  80
      L2=K-9-L1*5                                                  81
      XL=KP(L1,L2)                                                 82
      GOTO 9                                                       83
   79 XL=ABS(X(L,K))                                               84
      IF(K.EQ.3) GOTO 14                                           85
      IF(XL.GT.0.) GOTO 1                                          86
      LIM(I)=1.E-6                                                 87
      GOTO 73                                                      88
   14 LIM(I)=5.                                                    89
      GOTO 73                                                      90
    1 IF(K.NE.5) GOTO 9                                            91
      LIM(I)=0.5*XL                                                92
      GOTO 73                                                      93
    9 LIM(I)=0.1*XL                                                94
   73 TOLL(I)=0.0001*XL                                            95
      NCC=NCC+1                                                    96
   52 COH=X(L,1)                                                   97
      CTH=X(L,2)                                                   98
      EO=X(L,3)                                                    99
      KW=X(L,4)                                                   100
      JA=X(L,5)                                                   101
      SL=X(L,6)/AL10                                              102
      VO=VVO(L)                                                   103
      FDILH=FDILHM(L)                                             104
      K2=LAST(L)                                                  105
      DO 26 K=K1,K2                                               106
      IF(EO.GT.0)GO TO 16                                         107
      PHO=-E(K)*AL10                                              108
      GO TO 21                                                    109
   16 PHO=(E(K)-EO)/SL                                            110
   21 H=EXP(PHO)                                                  111
      DHEO=-H/SL                                                  112
      DHJA=0.                                                     113
      DHSL=-H*PHO/SL                                              114
      IF(JA.EQ.0.) GOTO 89                                        115
C CALCULATES H VALUE AND THE DERIVATIVES OF H WITH RESPECT TO EO, JA, SL   116
      CALL HH(JA,PHO,SL,H,JAV,JEO,JSL,DHEO,DHJA,DHSL)             117
   89 HQ=H*H                                                      118
      SAN1CO=0.                                                   119
      SAN2CT=0.                                                   120
      SDAHCO=0.                                                   121
      SDAHCT=0.                                                   122
      IF(NL.EQ.0) GOTO 24                                         123
      DO 29 J=1,NL                                                124
      I=J*2+5                                                     125
      COL(J)=X(L,I)                                               126
      CTL(J)=X(L,I+1)                                             127
      NPO(J)=NNPO(L,J)                                            128
      NPT(J)=NNPT(L,J)                                            129
      FDILL(J)=FDILLM(L,J)                                        130
      A1=0.                                                       131
      DA1=0.                                                      132
      DA2=0.                                                      133
```

```
            A2=1.                                                            134
            NBL=NB(J)                                                        135
            DO 30 I=1,NBL                                                    136
            BH=B(J,I)*H**I                                                   137
            DA1=DA1+I*I*BH/H                                                 138
            DA2=DA2+I*BH/H                                                   139
            A1=A1+I*BH                                                       140
         30 A2=A2+BH                                                         141
            A2Q=A2*A2                                                        142
            A1M(J)=A1                                                        143
            A2M(J)=A2                                                        144
            AM(J)=A1/A2                                                      145
            DAH(J)=(DA1*A2-DA2*A1)/A2Q                                       146
            SAN1CO=SAN1CO+(AM(J)-NPO(J))*COL(J)*FDILL(J)                     147
            SAN2CT=SAN2CT+(AM(J)-NPT(J))*CTL(J)                              148
            SDAHCO=SDAHCO+DAH(J)*COL(J)*FDILL(J)                             149
         29 SDAHCT=SDAHCT+DAH(J)*CTL(J)                                      150
         24 NUM=VO*(SAN1CO+H-KW/H-COH*FDILH)                                 151
            DEN=KW/H-H-SAN2CT+CTH                                            152
            DENQ=DEN*DEN                                                     153
            VC=NUM/DEN                                                       154
            DV(K)=VC-V(K)                                                    155
            IF(JVH.EQ.0) GOTO 6                                              156
            DVH=(VO*(SDAHCO+1.+KW/HQ)*DEN+(KW/HQ+1.+SDAHCT)*NUM)/DENQ        157
          6 IF(ISDV.EQ.1) GOTO 26                                            158
            IF(NCG.LE.NC.AND.NCG.GT.0) GOTO 60                               159
            IF(E0.GT.0.)GO TO 97                                             160
            PH(K)=E(K)                                                       161
            GO TO 98                                                         162
         97 PH(K)=-ALOG(H)/AL10                                              163
         98 IF(NL.EQ.0)GO TO 26                                              164
            DO 18 J=1,NL                                                     165
         18 Z(K,J)=AM(J)                                                     166
            GOTO 26                                                          167
         60 IF(JB*LEV.NE.1) GOTO 7                                           168
            DO 17 IK=1,NL                                                    169
            NBL=NB(IK)                                                       170
            A2K(IK,NBL)=B(IK,NBL)*H**NBL                                     171
            A1K(IK,NBL)=NBL*A2K(IK,NBL)                                      172
            IF(NBL.EQ.1) GOTO 17                                             173
            DO 58 I=2,NBL                                                    174
            J=NBL+1-I                                                        175
            BH=B(IK,J)*H**J                                                  176
            A1K(IK,J)=A1K(IK,J+1)+J*BH                                       177
         58 A2K(IK,J)=A2K(IK,J+1)+BH                                         178
         17 CONTINUE                                                         179
C     CALCULATES THE DERIVATIVES OF V WITH RESPECT TO THE PARAMETERS         180
C     TO BE REFINED                                                          181
          7 CALL DERIV(NV,IVAR,VO,FDILH,DEN,NUM,DENQ,DHEQ,H,DVH,DHJA,DHSL,   182
           &AM,NPO,FDILL,NPT,COL,CTL,A1K,A2M,A2K,A1M,KP,DER)                 183
            DO 33 I=1,NV                                                     184
         33 D(K,I)=DER(I)                                                    185
         26 CONTINUE                                                         186
            IF(LEV.EQ.2) GOTO 39                                             187
            IF(L.EQ.NTIT) GOTO 15                                            188
            L=L+1                                                            189
            K1=K2+1                                                          190
            GOTO 52                                                          191
C     CALCULATES STANDARD DEVIATIONS AND R(HAMILTON)                         192
         15 K1=1                                                             193
         39 SDV2=0.                                                          194
            DO 41 K=K1,K2                                                    195
         41 SDV2=SDV2+DV(K)*DV(K)                                            196
            R=SQRT(SDV2/SV2)                                                 197
            SDV=SQRT(SDV2/(N-NV))                                            198
            IF(ISDV.EQ.1) GOTO 56                                            199
            IF(NCG.EQ.0) GOTO 80                                             200
            IF(NCG*LEV*NCC.EQ.1) WRITE(6,105) SDV,R                          201
        105 FORMAT(/* ST.DEV.=*E10.3,5X*R(HAMILTON)=*E10.3,5X*WITH THE INPUT D
           &ATA*/)                                                           203
C     CALCULATES THE SHIFTS                                                  204
            CALL SHIFT(NV,K1,K2,DV,D,SH,SIGX)                                205
C     COMPARES THE SHIFTS WITH LIMITS AND TOLERANCES                         206
            M=0                                                              207
            DO 40 I=1,NV                                                     208
            IF(ABS(SH(I)).LE.TOLL(I)) GOTO 40                                209
            M=1                                                              210
            MLEV=1                                                           211
```

```
           IF(SH(I).LT.(-LIM(I))) SH(I)=-LIM(I)                                 212
           IF(SH(I).GT.LIM(I)) SH(I)=LIM(I)                                     213
        40 CONTINUE                                                             214
           IF(M.EQ.1.AND.NCC.LT.NCICP) GOTO 91                                  215
           NCIC=NCC                                                             216
           NCC=NCICP                                                            217
           IF(NCIC.LT.NCICP) GOTO 91                                            218
C    IF THE CYCLE DOES NOT CONVERGE, ASSIGNS TO THE PARAMETERS THE VALUE OF     219
C    THE PREVIOUS CYCLE                                                         220
           IF(LEV.EQ.2) GOTO 67                                                 221
           DO 66 K=1,NV                                                         222
           J=IVAR(K)                                                            223
           IF(J.LT.15) GOTO 11                                                  224
           L1=(J-10)/5                                                          225
           L2=J-9-L1*5                                                          226
           KP(L1,L2)=XM(K)                                                      227
           B(L1,L2)=BM(L1,L2)                                                   228
           GOTO 66                                                              229
        11 DO 96 L=1,NTIT                                                       230
        96 X(L,J)=XM(K)                                                         231
        66 SIGT(J)=0.                                                           232
           GOTO 8                                                               233
        67 DO 68 K=1,NV                                                         234
           J=IVAR(K)                                                            235
           SIGP(L,J)=0.                                                         236
           DEVS(L)=0.                                                           237
           RL(L)=0.                                                             238
        68 X(L,J)=XM(K)                                                         239
           GOTO 83                                                              240
        91 IF(NCG.EQ.1.OR.LEV.EQ.2) GOTO 3                                      241
           IF(SDV2.LE.SDV2M) GOTO 3                                             242
           DO 2 I=1,NV                                                          243
         2 SH(I)=SH(I)/3.                                                       244
     GIVES THE PARAMETERS AN INCREMENT EQUAL TO THE CALCULATED SHIFT            245
         3 DO 85 I=1,NV                                                         246
           K=IVAR(I)                                                            247
           XX=SH(I)                                                             248
           IF(K.LT.15) GOTO 19                                                  249
           L1=(K-10)/5                                                          250
           L2=K-9-L1*5                                                          251
           KP(L1,L2)=KP(L1,L2)+XX                                               252
           XV=KP(L1,L2)                                                         253
        65 SIGX(I)=SIGX(I)/(AL10*XV)                                            254
           XV=ALOG(XV)/AL10                                                     255
           GOTO 50                                                              256
        19 X(L,K)=X(L,K)+XX                                                     257
           XV=X(L,K)                                                            258
           IF(LEV.EQ.2) GOTO 88                                                 259
           DO 77 J=1,NTIT                                                       260
        77 X(J,K)=XV                                                            261
        88 IF(K.EQ.4) GOTO 65                                                   262
        50 IF(LEV.EQ.1) GOTO 62                                                 263
           XPP(L,K)=XV                                                          264
           SIGP(L,K)=SIGX(I)                                                    265
           GOTO 85                                                              266
        62 XPT(K)=XV                                                            267
           SIGT(K)=SIGX(I)                                                      268
        85 CONTINUE                                                             269
           IF(LEV.EQ.2) GOTO 59                                                 270
           IF(JB.EQ.0) GOTO 38                                                  271
           DO 57 K=1,NL                                                         272
           NBL=NB(K)                                                            273
           BL=1.                                                                274
           DO 57 I=1,NBL                                                        275
           BL=BL*KP(K,I)                                                        276
        57 B(K,I)=BL                                                            277
           GOTO 38                                                              278
        59 IF(M.EQ.1) GOTO 86                                                   279
           ISDV=1                                                               280
           GOTO 52                                                              281
        83 IF(L.EQ.NTIT) GOTO 84                                                282
           L=L+1                                                                283
           K1=K2+1                                                              284
           K2=LAST(L)                                                           285
           GOTO 35                                                              286
        38 IF(M.EQ.1) GOTO 81                                                   287
         8 IF(NCIC.EQ.1.AND.NCG.GT.1) GOTO 28                                   288
           ISDV=1                                                               289
```

```
              GOTO 54                                                    290
        84 LEV=1                                                         291
           IF(MLEV.EQ.0.OR.NCG.EQ.NC) GOTO 28                            292
           NCG=NCG+1                                                     293
           GOTO 82                                                       294
        22 NCGP=NCIC                                                     295
           NC=NCICP                                                      296
           GO TO 23                                                      297
        28 NCGP=NCG                                                      298
        23 NCG=0                                                         299
           GOTO 81                                                       300
        56 ISDV=0                                                        301
           IF(LEV2.EQ.0) GOTO 22                                         302
           IF(LEV.EQ.1) GOTO 12                                          303
           DO 78 I=1,NV                                                  304
           K=IVAR(I)                                                     305
        78 SIGP(L,K)=SIGP(L,K)*SDV                                       306
           DEVS(L)=SDV                                                   307
           RL(L)=R                                                       308
           GOTO 83                                                       309
        12 IF(NCG.EQ.1) GOTO 99                                          310
           IF(ABS(SDV2-SDV2M)/SDV2.LE.0.0001) GOTO 28                    311
        99 SDV2M=SDV2                                                    312
           IF(LEV2.EQ.0) GOTO 28                                         313
           LEV=2                                                         314
           GOTO 81                                                       315
        80 CALL OUT(NTIT,NVT,DEVS,RL,IVT,SIGP,XPP,NCGP,NC,SDV,R,NVAR,KVAR, 316
          &SIGT,XPT,NL,LAST,NTP,V,DV,E,PH,Z)                             317
           GOTO 70                                                       318
           END                                                           319
           SUBROUTINE OUT(NTIT,NVT,DEVS,RL,IVT,SIGP,XPP,NCGP,NC,SDV,R,NVAR, 320
          &KVAR,SIGT,XPT,NL,LAST,NTP,V,DV,E,PH,Z)                        321
C                                                                        322
C    THIS ROUTINE PRINTS OUT THE RESULTS                                 323
C                                                                        324
           DIMENSION NVT(20),DEVS(20),RL(20),IVT(20,6),SIGP(20,14),XPP(20,14), 325
          &VAR(34),KVAR(12),SIGT(34),XPT(34),V(700),DV(700),E(700),PH(700), 326
          &Z(700,4),LAST(12)                                             327
           DATA VAR/3HCOH,3HCTH,2HEO,3HKWL,2HJA,2HSL,4HCOL1,4HCTL1,4HCOL2,4HC 328
          &TL2,4HCOL3,4HCTL3,4HCOL4,4HCTL4,3HK11,3HK12,3HK13,3HK14,3HK15,3HK2 329
          &1,3HK22,3HK23,3HK24,3HK25,3HK31,3HK32,3HK33,3HK34,3HK35,3HK41,3HK4 330
          &2,3HK43,3HK44,3HK45/                                          331
           DATA VARZ/1HZ/                                                332
           DO 64 L=1,NTIT                                                333
           N=NVT(L)                                                      334
           IF(N.EQ.0) GOTO 64                                            335
           WRITE(6,113) L,DEVS(L),RL(L)                                  336
       113 FORMAT(//* TITRATION*I3,5X*ST.DEV.=*E10.3,5X*R(HAMILTON)=*E10.3) 337
           WRITE(6,116)                                                  338
       116 FORMAT(/28X*VALUE*13X*ST.DEV.*/)                              339
           DO 23 I=1,N                                                   340
           K=IVT(L,I)                                                    341
        23 WRITE(6,115) VAR(K),XPP(L,K),SIGP(L,K)                        342
       115 FORMAT(17X,A4,F15.7,5X,E14.7)                                 343
        64 CONTINUE                                                      344
           IF(NCGP.GE.NC) WRITE(6,117)                                   345
       117 FORMAT(/* MAXIMUM NO. OF ITERATIONS PERFORMED*)               346
           WRITE(6,107) NCGP,SDV,R                                       347
       107 FORMAT(/* CYCLE N.*I3,7X*ST.DEV.=*E10.3,5X*R(HAMILTON)=*E10.3) 348
           WRITE(6,116)                                                  349
           DO 16 I=1,NVAR                                                350
           K=KVAR(I)                                                     351
           SIGT(K)=SIGT(K)*SDV                                           352
        16 WRITE(6,115) VAR(K),XPT(K),SIGT(K)                            353
           WRITE(6,121)                                                  354
       121 FORMAT(/11X*V*6X*DV*6X*E*6X*PH*)                              355
           IF(NL.GT.0) WRITE(6,123) (VARZ,K,K=1,NL)                      356
       123 FORMAT(1H+35X4(4XA1,I1))                                      357
           WRITE(6,124)                                                  358
       124 FORMAT(1H+/)                                                  359
           L=1                                                           360
           L1=1                                                          361
           DO 14 K=1,NTP                                                 362
           WRITE(6,122) V(K),DV(K),PH(K)                                 363
       122 FORMAT(5X,F9.4,F7.4,8X,F7.3)                                  364
           IF(E(K).NE.PH(K))WRITE(6,125)E(K)                             365
       125 FORMAT(1H+20XF8.2)                                            366
           IF(K.LT.L1) GOTO 10                                           367
```

Appendix B: ACBA

```
      WRITE(6,120) L
 120  FORMAT(1H+,I4)
      L1=LAST(L)+1
      L=L+1
  10  IF(NL.GT.0) WRITE(6,118) (Z(K,J),J=1,NL)
 118  FORMAT(1H+,36X4F6.3)
  14  CONTINUE
      RETURN
      END
      SUBROUTINE SHIFT(NV,K1,K2,DV,D,SH,SIGX)
C
C  THIS ROUTINE CALCULATES THE MATRIX OF THE COEFFICIENTS IN ORDER TO
C  DETERMINE THE SHIFTS
C
      DIMENSION CK(12),C(12,12),DV(700),D(700,12),SH(12),SIGX(12)
     &,QAST(12,12),BAST(12),VAL(12),VEC(12,12),XAST(12),DIAG(12)
      DO 22 I=1,NV
      CK(I)=0.
      DO 22 J=1,NV
  22  C(I,J)=0.
      DO 31 K=K1,K2
      DO 31 I=1,NV
      CK(I)=CK(I)-DV(K)*D(K,I)
      DO 31 J=1,NV
  31  C(I,J)=C(I,J)+D(K,I)*D(K,J)
      IF(NV.GT.1) GOTO 32
      SH(1)=CK(1)/C(1,1)
      SIGX(1)=1./SQRT(C(1,1))
      RETURN
C  RECALLS THE SUBROUTINES CONCERNING THE SHIFT CALCULATION
  32  CALL SCALE(NV,C,CK,QAST,BAST)
      CALL JACOBI(NV,QAST,VAL,VEC)
      CALL MIAVA(NV,VAL,VEC,BAST,XAST,DIAG)
C  CALCULATES SHIFTS AND RELATIVE STANDARD DEVIATIONS
      DO 15 I=1,NV
      SH(I)=XAST(I)/SQRT(C(I,I))
  15  SIGX(I)=SQRT(DIAG(I)/C(I,I))
      RETURN
      END
      SUBROUTINE DATI(NC,NL,NVAR,SVQ,KVAR,KP,B,NB,X,VV0,FDILHM,FDILLM,
     &NNPO,NNPT,NVT,IVT,LEV2,V,E,LAST,NP,NTP,NTIT,AL10,JB,SVQT)
C
C  THIS ROUTINE READS INPUT DATA
C
      DIMENSION TITLE(20),PVAR(12),VAR(34),KVAR(12),B(4,5),NB(4),VX(14),
     &NNPO(20,4),NNPT(20,4),FDILLM(20,4),X(20,14),VV0(20),FDILHM(20),
     &NVT(20),IVT(20,6),VV(5),EE(5),INDEX(5),V(700),E(700),LAST(20),
     &NP(20),SVQ(20),KE(700)
      REAL KL(4,5),KP(4,5),KWL
      DATA VAR/3HCOH,3HCTH,2HEO,3HKWL,2HJA,2HSL,4HCOL1,4HCTL1,4HCOL2,4HC
     &TL2,
     &4HCOL3,4HCTL3,4HCOL4,4HCTL4,3HK11,3HK12,3HK13,3HK14,3HK15,3HK21,
     &3HK22,3HK23,3HK24,3HK25,3HK31,3HK32,3HK33,3HK34,3HK35,3HK41,3HK42,
     &3HK43,3HK44,3HK45/
      AL10=ALOG(10.)
      READ(5,100) TITLE
 100  FORMAT(20A4)
      READ(5,101) NC,NL,NVAR,(PVAR(I),I=1,NVAR)
 101  FORMAT(3I2,12A5)
C
C  NC=NO. OF CYCLES    -    NL=NO. OF LIGANDS
C  NVAR=NO. OF COMMON PARAMETERS TO BE REFINED
C  PVAR=SYMBOLS OF COMMON PARAMETERS TO BE REFINED
C  (COH,CTH,EO,KWL,JA,SL,COL1,CTL1,COL2,CTL2,COL3,CTL3,COL4,CTL4,K11,K12,
C  K13,K14,K15,K21,K22,K23,K24,K25,K31,K32,K33,K34,K35,K41,K42,K43,K44,K45)
C
      IF(NVAR.EQ.0) STOP
      WRITE(6,102) TITLE
 102  FORMAT(1H1,10X20A4/)
      JB=0
      L1=1
      DO 69 I=1,NVAR
      DO 94 K=1,34
      IF(PVAR(I).EQ.VAR(K)) GOTO 20
  94  CONTINUE
  20  IF(K.GT.14) JB=1
  69  KVAR(I)=K
      IF(NL.EQ.0) GOTO 1
```

```
      READ(5,103) ((KL(I,K),K=1,5),I=1,NL)                          446
  103 FORMAT(5F10.0)                                                 447
C  KL= LOG(K)                                                        448
      DO 2 I=1,NL                                                    449
      BL=0.                                                          450
      DO 3 K=1,5                                                     451
      AKL=KL(I,K)                                                    452
      IF(AKL.EQ.0.) GOTO 2                                           453
      BL=BL+AKL                                                      454
      KP(I,K)=EXP(AL10*AKL)                                          455
      WRITE(6,104) I,K,AKL                                           456
  104 FORMAT(* LOG.K*2I1,*=*,F8.3)                                   457
    3 B(I,K)=EXP(AL10*BL)                                            458
    2 NB(I)=K-1                                                      459
    1 READ(5,103) TEMP,KWL                                           460
      VX(4)=EXP(AL10*KWL)                                            461
      VX(6)=0.086173*(273.15+TEMP)*AL10                              462
C  TEMP=TEMPERATURE  -  KWL=LOG(KW)  -  VX(4)=KW  -  VX(6)=SL        463
      WRITE(6,110) TEMP,VX(6),KWL                                    464
  110 FORMAT(/5X*TEMPERATURE=*F9.3/5X*SL=*F9.3/5X*KWL=*F8.3/)         465
      I=0                                                            466
      L=0                                                            467
      LEV2=0                                                         468
      SVQT=0.                                                        469
   75 READ(5,100) TITLE                                              470
      READ(5,112) VX(1),VX(2),VO,VX(3),VX(5),FDILH,VIN,DV,NA         471
  112 FORMAT(2F10.0,3F7.0,F10.0,2F7.0,I3)                            472
C                                                                    473
C  VX(1)=COH  -  VX(2)=CTH  -  VO=VOLUME  -  VX(3)=EO  -  VX(5)=JA   474
C  FDILH = DILUTION FACTOR FOR COH                                   475
C  VIN = VOLUME OF THE FIRST POINT OF THE TITRATION                  476
C  DV = INCREMENT OF THE VOLUME FOR THE FOLLOWING POINTS             477
C  NA = NUMBER OF POINTS IN THE TITRATION                            478
C                                                                    479
      IF(VO.EQ.0.) GOTO 76                                           480
      SV2=0.                                                         481
      L=L+1                                                          482
      WRITE(6,119) L,TITLE                                           483
  119 FORMAT(/* TITRATION*I3,5X20A4/)                                484
      IF(FDILH.EQ.0..AND.VX(1).NE.0.) FDILH=1.                       485
      WRITE(6,109) VX(1),FDILH,VX(2),VO,VX(3),VX(5)                  486
  109 FORMAT(6X*COH =*F11.8,4X*FDILH=*F11.8,3X*CTH =*F11.8,3X*VO=*F8.3, 487
     83X*EO=*F8.3,3X*JA=*F8.3)                                       488
      IF(NL.EQ.0) GOTO 61                                            489
      DO 98 K=1,NL                                                   490
      J=2*K+5                                                        491
      READ(5,210) VX(J),VX(J+1),NNPO(L,K),NNPT(L,K),FDILLM(L,K)      492
  210 FORMAT(2F10.0,2I2,F10.0)                                       493
C                                                                    494
C  VX(J)=COL  -  VX(J+1)=CTL                                         495
C  NNPO = NUMBER OF PROTONS IN THE LIGAND (TITRATE)                  496
C  NNPT = NUMBER OF PROTONS IN THE LIGAND (TITRANT)                  497
C  FDILLM = DILUTION FACTOR FOR COL                                  498
C                                                                    499
      IF(FDILLM(L,K).EQ.0..AND.VX(J).NE.0.) FDILLM(L,K)=1.           500
   98 WRITE(6,108)K,VX(J),K,FDILLM(L,K),K,VX(J+1),K,NNPO(L,K),K,NNPT(L,K) 501
  108 FORMAT(6X*COL*I1,*=*F11.8,4X*FDIL*I1,*=*F11.8,3X*CTL*I1,*=*F11.8, 502
     83X,*NPO*I1,*=*I2,3X*NPT*I1,*=*I2)                              503
   61 READ(5,123) NVART,(PVAR(K),K=1,NVART)                          504
  123 FORMAT(I2,12A5)                                                505
C  NVART=NO. OF PARAMETERS TO BE REFINED                             506
C  PVAR=SYMBOLS OF PARAMETERS TO BE REFINED                          507
      DO 74 J=1,14                                                   508
   74 X(L,J)=VX(J)                                                   509
      VVO(L)=VO                                                      510
      FDILHM(L)=FDILH                                                511
      NVT(L)=NVART                                                   512
      IF(NVART.EQ.0) GOTO 4                                          513
      LEV2=1                                                         514
      DO 95 K=1,NVART                                                515
      DO 97 J=1,14                                                   516
      IF(PVAR(K).EQ.VAR(J)) GOTO 95                                  517
   97 CONTINUE                                                       518
   95 IVT(L,K)=J                                                     519
    4 IF(DV.EQ.0.)GO TO 25                                           520
      READ(5,*)(KE(K),K=1,NA)                                        521
C                                                                    522
C  KE = EMF*10 (IF EO.GT.0) OR PH*1000 (IF EO=0)                     523
```

Appendix B: ACBA

```
C     IF (KE=0 AND EO=0) OR (KE=9999 AND EO.GT.0) THE PAIR OF VALUES V,KE
C     IS NEGLECTED
C
      DO 5 K=1,NA
      IF(VX(3).GT.0.) GO TO 8
      IF(KE(K).EQ.0) GO TO 6
      I=I+1
      E(I)=KE(K)*0.001
      GO TO 7
    8 IF(KE(K).EQ.9999) GO TO 6
      I=I+1
      E(I)=KE(K)*0.1
    7 V(I)=VIN
      SV2=SV2+VIN*VIN
    6 VIN=VIN+DV
    5 CONTINUE
      GO TO 72
   25 READ(5,114) (VV(K),EE(K),INDEX(K),K=1,5)
  114 FORMAT(5(2F7.2,I1))
C
C     VV = VOLUME    -   EE = EMF (IF EO.GT.0)  OR  PH (IF EO = 0)
C     INDEX = 0 NORMAL  -   INDEX = 1 END OF TITRATION
C
      DO 71 K=1,5
      IF(VV(K).EQ.0..AND.EE(K).EQ.0.) GOTO 71
      I=I+1
      V(I)=VV(K)
      E(I)=EE(K)
      SV2=SV2+V(I)*V(I)
      IF(INDEX(K).EQ.1) GOTO 72
   71 CONTINUE
      GOTO 25
   72 N=I-L1+1
      WRITE(6,111) N
  111 FORMAT(/6X*NO. OF POINTS*I4/)
      L1=I+1
      LAST(L)=I
      NP(L)=N
      SVQ(L)=SV2
      SVQT=SVQT+SV2
      GOTO 75
   76 NTP=I
      WRITE(6,120) NTP
  120 FORMAT(/* TOTAL NO. OF POINTS*I4/)
      NTIT=L
      RETURN
      END
      SUBROUTINE HH(JA,PHO,SL,H,JAV,JEO,JSL,DHEO,DHJA,DHSL)
C
C     THIS ROUTINE CALCULATES THE DERIVATIVES OF H WITH RESPECT TO EO, JA, SL.
C     CALCULATES ALSO H-VALUE.
C
      REAL JA
      PSI=PHO
   63 PH1=PSI-JA*H/SL
      H=EXP(PH1)
      IF(JAV.EQ.0) GOTO 5
      DHJA=-H*(H+JA*DHJA)/SL
    5 IF(JEO.EQ.0) GOTO 75
      DHEO=-H*(1.+JA*DHEO)/SL
   75 IF(JSL.EQ.0) GOTO 3
      DHSL=H*(PSI*SL+JA*H)/(SL*SL)
    3 IF(ABS(PHO-PH1)/PH1.LT.0.0001) RETURN
      PHO=PH1
      GOTO 63
      END
      SUBROUTINE DERIV(NV,IVAR,VO,FDILH,DEN,NUM,DENQ,DHEO,H,DVH,
     &DHJA,DHSL,AM,NPO,FDILL,NPT,COL,CTL,A1K,A2M,A2K,A1M,KP,DER)
C
C     THIS ROUTINE CALCULATES THE DERIVATIVES OF V WITH RESPECT TO THE
C     PARAMETERS TO BE REFINED
C
      DIMENSION IVAR(12),AM(4),NPO(4),FDILL(4),NPT(4),COL(4),CTL(4)
     &,A1K(4,5),A2M(4),A2K(4,5),A1M(4),DER(12)
      REAL KP(4,5),NUM
      DO 33 I=1,NV
      J=IVAR(I)
      IF(J.GT.14) GOTO 49
```

```
      GOTO(41,42,43,44,45,46,47,48,47,48,47,48,47,48) J            602
   41 DERV=-VO*FDILH/DEN                                            603
      GOTO 33                                                       604
   42 DERV=-NUM/DENQ                                                605
      GOTO 33                                                       606
   43 DERV=DVH*DHEO                                                 607
      GOTO 33                                                       608
   44 DERV=-(VO*DEN+NUM)/(H*DENQ)                                   609
      GOTO 33                                                       610
   45 DERV=DVH*DHJA                                                 611
      GOTO 33                                                       612
   46 DERV=DVH*DHSL                                                 613
      GOTO 33                                                       614
   47 L1=(J-5)/2                                                    615
      DERV=VO*(AM(L1)-NPO(L1))*FDILL(L1)/DEN                        616
      GOTO 33                                                       617
   48 L1=(J-6)/2                                                    618
      DERV=NUM*(AM(L1)-NPT(L1))/DENQ                                619
      GOTO 33                                                       620
   49 L1=(J-10)/5                                                   621
      L2=J-9-L1*5                                                   622
      DVA=(VO*COL(L1)*FDILL(L1)*DEN+CTL(L1)*NUM)/DENQ               623
      DERV=DVA*(A1K(L1,L2)*A2M(L1)-A2K(L1,L2)*A1M(L1))/(KP(L1,L2)*A2M(L1  624
     8)**2)                                                         625
   33 DER(I)=DERV                                                   626
      RETURN                                                        627
      END                                                           628
      SUBROUTINE SCALE(N,Q,B,QAST,BAST)                             629
C                                                                   630
C     THIS ROUTINE SCALES THE MATRIX Q AND THE VECTOR B             631
C                                                                   632
      DIMENSION Q(12,12),QAST(12,12),BAST(12),B(12)                 633
      DO 2 I=1,N                                                    634
      QI=ABS(Q(I,I))                                                635
      BAST(I)=B(I)/SQRT(QI)                                         636
      QAST(I,I)=1.                                                  637
      DO 3 K=1,N                                                    638
      IF(I.EQ.K) GOTO 3                                             639
      QAST(I,K)=Q(I,K)/SQRT(QI*ABS(Q(K,K)))                         640
    3 CONTINUE                                                      641
    2 CONTINUE                                                      642
      RETURN                                                        643
      END                                                           644
      SUBROUTINE JACOBI(N,A,VAL,S)                                  645
C                                                                   646
C     THIS ROUTINE CALCULATES EIGENVALUES AND EIGENVECTORS OF THE SCALED  647
C     MATRIX A                                                      648
C                                                                   649
      DIMENSION A(12,12),S(12,12),VAL(12)                           650
      DO 42 I=1,N                                                   651
      DO 43 J=1,N                                                   652
   43 S(I,J)=0.                                                     653
   42 S(I,I)=1.                                                     654
      FN=0.                                                         655
      DO 1 I=2,N                                                    656
      I1=I-1                                                        657
      DO 2 J=1,I1                                                   658
    2 FN=FN+A(I,J)*A(I,J)                                           659
    1 CONTINUE                                                      660
      FN=SQRT(2.*FN)                                                661
      UN=FN*1.E-9                                                   662
      PN=N                                                          663
    3 FN=FN/PN                                                      664
    8 IN=0                                                          665
      DO 24 IQ=2,N                                                  666
      L=IQ-1                                                        667
      DO 22 IP=1,L                                                  668
      IF(ABS(A(IP,IQ)).LE.FN) GOTO 22                               669
      IN=1                                                          670
      V=-A(IP,IQ)                                                   671
      U=.5*(A(IP,IP)-A(IQ,IQ))                                      672
      W=SIGN(1.,U)*(V/SQRT(V*V+U*U))                                673
      DET=1.-W*W                                                    674
      SN=W/SQRT(2.*(1.+SQRT(DET)))                                  675
      CN=SQRT(1.-SN*SN)                                             676
      DO 15 I=1,N                                                   677
      BIP=A(I,IP)*CN-A(I,IQ)*SN                                     678
      BIQ=A(I,IP)*SN+A(I,IQ)*CN                                     679
```

Appendix B: ACBA

```
      A(I,IP)=BIP                                    680
      A(I,IQ)=BIQ                                    681
      BIP=S(I,IP)*CN-S(I,IQ)*SN                      682
      BIQ=S(I,IP)*SN+S(I,IQ)*CN                      683
      S(I,IP)=BIP                                    684
   15 S(I,IQ)=BIQ                                    685
      BIP=A(IP,IP)*CN-A(IQ,IP)*SN                    686
      BIQ=A(IP,IQ)*SN+A(IQ,IQ)*CN                    687
      BPQ=A(IP,IQ)*CN-A(IQ,IQ)*SN                    688
      BQP=A(IP,IP)*SN+A(IQ,IP)*CN                    689
      A(IP,IP)=BIP                                   690
      A(IQ,IQ)=BIQ                                   691
      A(IP,IQ)=BPQ                                   692
      A(IQ,IP)=BQP                                   693
      DO 20 I=1,N                                    694
      A(IP,I)=A(I,IP)                                695
   20 A(IQ,I)=A(I,IQ)                                696
   22 CONTINUE                                       697
   24 CONTINUE                                       698
      IF(IN.EQ.1) GOTO 8                             699
      IF(FN.GT.UN) GOTO 3                            700
      N1=N-1                                         701
      DO 30 K=1,N1                                   702
      L=0                                            703
      BIGA=A(K,K)                                    704
      K1=K+1                                         705
      DO 31 J=K1,N                                   706
      IF(BIGA.GE.A(J,J)) GOTO 31                     707
      BIGA=A(J,J)                                    708
      L=J                                            709
   31 CONTINUE                                       710
      IF(L.EQ.0) GOTO 30                             711
      A(L,L)=A(K,K)                                  712
      A(K,K)=BIGA                                    713
      DO 35 I=1,N                                    714
      SB=S(I,L)                                      715
      S(I,L)=S(I,K)                                  716
   35 S(I,K)=SB                                      717
   30 CONTINUE                                       718
      DO 50 I=1,N                                    719
   50 VAL(I)=A(I,I)                                  720
      RETURN                                         721
      END                                            722
      SUBROUTINE MIAVA(N,VAL,VEC,BAST,XAST,DIAG)     723
C                                                    724
C     THIS ROUTINE CALCULATES THE SCALED SHIFTS      725
C                                                    726
      DIMENSION VAL(12),VEC(12,12),BAST(12),XAST(12),DIAG(12)  727
      DO 1 I=1,N                                     728
      IF(ABS(VAL(I)).LE.1.E-6) VAL(I)=1.             729
    1 CONTINUE                                       730
      DO 4 I=1,N                                     731
      XAST(I)=0.                                     732
      DO 2 J=1,N                                     733
      S=0.                                           734
      DO 3 K=1,N                                     735
    3 S=S+VEC(I,K)*VEC(J,K)/ABS(VAL(K))              736
      IF(I.EQ.J) DIAG(I)=S                           737
    2 XAST(I)=XAST(I)+S*BAST(J)                      738
    4 CONTINUE                                       739
      RETURN                                         740
      END                                            741
```

Analysis using glass electrodes

Example 1. Titration of a weak acid with a weak base. Titrand: 0.6060M acetic acid (15 ml + 95 ml of 0.16M $NaClO_4$; dilution factor 0.13636). Titrant: 5.081M pyridine (input value 5). Parameters to be refined (i) $E°$ (EO), (ii) concentration of titrant (i.e. pyridine) (CTL2), (iii) log K_1^{II} for acetic acid (K11), (iv) log K_1^H for pyridine (K21).

```
    Input
        PYRIDINE-CH3COOH   I=0.15(NACLO4)
        99 2 4EO    CTL2 K11    K21
        4.60
        5.30
        25.        -13.69
            PYRIDINE-CH3COOH   I=0.15(NACLO4)
        0.         0.          110.   440.    0.      0.
        0.60597    0.          1 0 0.13636
        0.         5.          0 0 0.

        0.21  227.0  00.50  201.8  00.80  185.9  01.00  177.8  01.20  170.6  0
        1.50  161.4  01.80  153.6  02.09  147.1  02.45  140.3  02.80  134.8  0
        3.32  128.2  03.90  122.2  04.40  117.9  04.85  114.5  1
```

Output

```
            PYRIDINE-CH3COOH   I=0.15(NACLO4)
LOG.K11=   4.600
LOG.K21=   5.300

    TEMPERATURE=   25.000
    SL=    59.159
    KWL=  -13.690

TITRATION  1         PYRIDINE-CH3COOH   I=0.15(NACLO4)
    COH = 0.             FDILH= 0.            CTH  = 0.            VO= 110.000   EO= 440.000    JA=   0.
    COL1= 0.60597000     FDIL1= 0.13636000    CTL1= 0.             NPO1= 1       NPT1= 0
    COL2= 0.             FDIL2= 0.            CTL2= 5.00000000     NPO2= 0       NPT2= 0

    NO. OF POINTS  14

TOTAL NO. OF POINTS  14

ST.DEV.= 0.414E 00     R(HAMILTON)= 0.133E 00     WITH THE INPUT DATA
CYCLE N.  4     ST.DEV.= 0.471E-02     R(HAMILTON)= 0.152E-02
                        VALUE           ST.DEV.
        EO         442.8560295     0.1335093E 00
        CTL2         5.0778247     0.4476654E-01
        K11          4.5174921     0.4945248E-02
        K21          5.2582256     0.2922945E-02

         V       DV        E       PH      Z1     Z2
    1  0.2100  0.0036   227.00   3.649   0.881  0.976
       0.5000  0.0037   201.80   4.075   0.735  0.938
       0.8000  0.0043   185.90   4.343   0.599  0.892
       1.0000 -0.0011   177.80   4.480   0.521  0.857
       1.2000 -0.0020   170.60   4.602   0.451  0.819
       1.5000 -0.0054   161.40   4.758   0.365  0.760
       1.8000 -0.0065   153.60   4.889   0.298  0.700
       2.0900 -0.0026   147.10   4.999   0.248  0.645
       2.4500  0.0037   140.30   5.114   0.202  0.582
       2.8000  0.0078   134.80   5.207   0.170  0.529
       3.3200  0.0027   128.20   5.319   0.136  0.465
       3.9000  0.0010   122.20   5.420   0.111  0.408
       4.4000 -0.0027   117.90   5.493   0.096  0.368
       4.8500 -0.0020   114.50   5.550   0.085  0.338
```

Example 2. Titration of a mixture of weak bases with a strong acid. Titrand: 1.543mM pyridine + 1.872mM 2,2'-bipyridyl (input values 1 and 2). Titrant: 0.0997M $HClO_4$. Parameters to be refined (i) $E°$ (EO), (ii) concentration of pyridine (COL1), (iii) concentration of 2,2'-bipyridyl (COL2). In the output, Z_i is equal to the average number of protons bound to the ligand i.

```
    Input
        PYRIDINE-2,2-BIPYRIDYL   I=0.1(NACLO4)
        99 2 3EO    COL1 COL2
        5.33
        4.461
        25.        -13.75
            PYRIDINE-2,2-BIPYRIDYL   I=0.1(NACLO4)
        0.         0.0997      100.   420.   -480.    0.
        0.001      0.          0 0 0.
        0.002      0.          0 0 0.

        0.05   8.60   00.15  38.30  00.25  53.00  00.35  63.00  00.45  71.00  0
        0.60  80.60   00.75  88.90  00.90  96.10  01.04  102.2  01.15  106.7  0
        1.30  113.0   01.50  120.5  01.65  126.4  01.80  132.1  02.00  139.2  0
        2.15  144.9   02.30  150.5  02.46  156.9  02.65  164.4  02.81  171.6  0
        2.99  179.8   03.12  186.4  03.25  193.3  03.45  203.9  03.56  209.6  0
        3.70  216.1   03.85  221.9  04.00  227.2  04.15  231.8  04.40  238.0  1
```

Appendix B: ACBA

Output

```
              PYRIDINE-2,2-BIPYRIDYL  I=0.1(NACLO4)
LOG.K11=   5.330
LOG.K21=   4.461

   TEMPERATURE=    25.000
     SL=   59.159
     KWL=  -13.750

TITRATION  1        PYRIDINE-2,2-BIPYRIDYL I=0.1(NACLO4)
   COH = 0.              FDILH= 0.             CTH = 0.09970000    V0= 100.000    E0= 420.000    JA=-480.000
   COL1= 0.00100000      FDIL1= 1.00000000     CTL1= 0.            NP01= 0    NPT1= 0
   COL2= 0.00200000      FDIL2= 1.00000000     CTL2= 0.            NP02= 0    NPT2= 0

   NO. OF POINTS    30

TOTAL NO. OF POINTS    30

ST.DEV.= 0.456E 00     R(HAMILTON)= 0.175E 00      WITH THE INPUT DATA

CYCLE N.   6        ST.DEV.= 0.244E-02      R(HAMILTON)= 0.935E-03

                         VALUE              ST.DEV.
              E0       415.9951820        0.7832945E-01
              COL1       0.0015484        0.3835969E-05
              COL2       0.0018658        0.2824248E-05

        V         DV         E         PH        Z1      Z2
 1    0.0500  -0.0010      8.60      6.886     0.027   0.004
      0.1500  -0.0017     38.30      6.384     0.081   0.012
      0.2500  -0.0006     53.00      6.136     0.135   0.021
      0.3500  -0.0011     63.00      5.967     0.187   0.030
      0.4500  -0.0000     71.00      5.832     0.240   0.041
      0.6000  -0.0011     80.60      5.669     0.314   0.058
      0.7500   0.0020     88.90      5.529     0.387   0.079
      0.9000   0.0017     96.10      5.407     0.456   0.102
      1.0400  -0.0006    102.20      5.304     0.515   0.125
      1.1500  -0.0036    106.70      5.228     0.558   0.146
      1.3000   0.0025    113.00      5.122     0.618   0.179
      1.5000   0.0040    120.50      4.995     0.684   0.226
      1.6500   0.0023    126.40      4.895     0.731   0.269
      1.8000   0.0054    132.10      4.799     0.773   0.315
      2.0000  -0.0029    139.20      4.679     0.818   0.377
      2.1500   0.0003    144.90      4.582     0.848   0.431
      2.3000  -0.0016    150.50      4.488     0.874   0.485
      2.4600   0.0030    156.90      4.379     0.899   0.547
      2.6500  -0.0030    164.40      4.252     0.923   0.618
      2.8100   0.0026    171.60      4.131     0.941   0.682
      2.9900  -0.0026    179.80      3.992     0.956   0.747
      3.1200  -0.0013    186.40      3.880     0.966   0.792
      3.2500  -0.0002    193.30      3.763     0.974   0.833
      3.4500  -0.0007    203.90      3.583     0.982   0.883
      3.5600   0.0023    209.60      3.486     0.986   0.904
      3.7000   0.0035    216.10      3.376     0.989   0.924
      3.8500  -0.0032    221.90      3.277     0.991   0.939
      4.0000  -0.0014    227.20      3.186     0.993   0.950
      4.1500   0.0018    231.80      3.107     0.994   0.958
      4.4000  -0.0002    238.00      3.001     0.995   0.967
```

Appendix C. The MINIPOT program (Gaizer and Puskas 1981)

Program listing

The following is a listing of the MINIPOT program.

```
10 REM PROGRAMME MINIPOT
20 DIM L(40),M(40),N(40),E(40),P(6),T(6),A(50),B(50),E1(40)
30 DIM C(4,4),C5(4,4),D(40),G(4,40),S1(4),G1(4),G2(4),F(40)
40 DIM C1(40),C2(40),C3(40),C4(40),V1(40)
50 Z=LOG(10):Z1=1/Z:E5=1000
60 READ W:IF W>1 THEN 70:READ N1,N2,R,V0,M1,M2,M3,M4,M5,M6:GOTO 120
70 READ N1,N2,N3,N4,R,V0,M1,M2,M3,M4,M5,M6,L,J1,J2,J3
80 REM *N1,N2,N3* NO. OF MEASD. POINTS, PARAMS, COMPLEXES
90 REM *N4* PERCENT FOR T-M (=1) OR T-L (=2); *R* PRINT; *V0* VOL. OF TITRAN)
100 REM *M1,...,M6* T-H/M, T-OH, T-L IN TITRAND, THEN THOSE IN TITRAND
110 REM *L* START VAL. FOR FREE LIGAND; *J1,J2,J3* LOG(J-H, J-OH, K-S)
120 FOR I=1 TO N1:IF W>1 THEN 130:READ V1(I),E(I):GOTO 150
130 READ V1(I),E(I),B(I),A(I):B(I)=EXP(Z*B(I)):A(I)=EXP(Z*A(I))
140 REM *V1* VOL. OF TITRANT; *E* MEASD. EMF; *B,A* LOG START VALUE
        FOR FREE METAL/H AND LIGAND;
150 X1=V0+V1(I):M(I)=(V0*M1+V1(I)*M4)/X1:N(I)=(V0*M2+V1(I)*M5)/X1
160 IF W=1 THEN 170:L(I)=(V0*M3+V1(I)*M6)/X1
170 IF R=0 THEN 180:PRINTUSING 1340,I;M(I)*E5;N(I)*E5;L(I)*E5;E(I);V1(I)
180 NEXT I:PRINT :PRINT
190 IF W=1 THEN 210:READ Q1,P1,Q2,P2,Q3,P3,Q4,P4
200 REM *Q1,P1,...,Q4,P4* STOICH. COEFFS. OF SPECIES
210 FOR I=1 TO N2:READ P(I),T(I):NEXT I
220 REM *P,T* PARAMETER AND INCREMENT. SEQUENCE, IF W=1, THEN LOG(K-S,
        J-H, J-OH, T-OH), E-0, G; IF W=2, THEN LOG(BETA-1,...,4),E-0, G
230 INPUT K,V,N:IF W=1 THEN 260:INPUT T1
240 REM *K,V,N* FIRST, LAST, EXP. POINT AND STEP;
250 REM *T1* IF THERE ARE ROOTS =-1, OTHERWISE =+1
260 INPUT F,R:GOSUB 830:GOSUB 1280:U1=U:IF F=0 THEN 410
270 REM *F* THE TASK TO BE EXECUTED; *R* PRINT
280 REM *** SEARCH PROCEDURE FOR PARAMETER VALUE ***
290 INPUT T3,T4,T5,T6,R,T7,V1:IF F=1 THEN 340
```

Appendix C: MINIPOT

```
300 REM *T3,T4,T5* START, STEP, UPPER LIMIT OF PARAM. SEARCH;
310 REM *T6* SER. NO. OF PARAM. TO BE SEARCHED FOR;
320 REM *R* PRINT; *T7* STEPS AFTER MINIMUM; *V1* CONTROL
330 T3=P(T6)*T3:T4=P(T6)*T4:T5=P(T6)*T5
340 P(T6)=T3:GOSUB 1470:GOSUB 830:GOSUB 1280:T8=H:T9=P(T6):GOTO 370
350 GOSUB 1470:GOSUB 830:GOSUB 1280:IF H>T8 THEN 360:T8=H:T9=P(T6):GOTO 370
360 T7=T7-1:IF T7=0 THEN 380
370 P(T6)=P(T6)+T4:IF P(T6)<=T5 THEN 350
380 P(T6)=T9:PRINTUSING 390,T6,T9:GOSUB 1470:ON V1 GOTO 60,190,230,260,410:END

390 %THE OPTIMUM VALUE OF PARAMETER NO. #### IS =-########.######
400 REM *** REFINEMENT PROCEDURE FOR PARAMETER VALUES ***
410 INPUT F1,F2,V1,R:MAT C=ZER(F1,F1):MAT C5=ZER(F1,F1)
420 REM *F1* NO. OF PARAM(S). TO BE REFINED; *F2* IF =1, THEN CORR. VECTOR
        OPT., OTHERWISE =0; *V1* CONTROL; *R* PRINT
430 MAT INPUT S1(F1):PRINT "PARAMETERS TO BE REFINED:":MAT PRINT S1;
440 REM *S1* SER. NO. OF PARAM(S). TO BE REFINED
450 REM *** ERROR VECTOR, D ***
460 FOR I=K TO V STEP N:D(I)=E(I)-E1(I):F(I)=E1(I):NEXT I
470 REM *** G MATRIX ***
480 FOR K1=1 TO F1:J=S1(K1):P(J)=P(J)+T(J):T6=J:GOSUB 1470:GOSUB 830
490 FOR I=K TO V STEP N:G(K1,I)=(E1(I)-F(I))/T(J):NEXT I
500 P(J)=P(J)-T(J):GOSUB 1470:NEXT K1
510 REM *** C MATRIX ***
520 FOR K1=1 TO F1:FOR K2=1 TO F1:C(K1,K2)=0:FOR I=K TO V STEP N
530 C(K1,K2)=C(K1,K2)+G(K1,I)*G(K2,I):NEXT I
540 C(K2,K1)=C(K1,K2):NEXT K2:NEXT K1
550 REM *** INVERSION OF MATRIX C INTO C5 ***
560 FOR I=1 TO F1:FOR K1=1 TO F1:C5(I,K1),C5(K1,I)=0:NEXT K1:C5(I,I)=1
570 NEXT I:C5(F1,F1)=1:FOR K1=1 TO F1:V3=C(K1,K1):FOR I=1 TO F1
580 C(K1,I)=C(K1,I)/V3:C5(K1,I)=C5(K1,I)/V3:NEXT I:FOR L=1 TO F1
590 IF L-K1=0 THEN 610:V3=C(L,K1):FOR I=1 TO F1
600 C(L,I)=C(L,I)-V3*C(K1,I):C5(L,I)=C5(L,I)-V3*C5(K1,I):NEXT I
610 NEXT L:NEXT K1
620 REM *** MULTIPL. OF ERROR VECT AND G-TRANSP. ***
630 FOR I=1 TO F1:S=0:FOR J=1 TO V S=S+G(I,J)*D(J):NEXT J:G1(I)=S:NEXT I
```

```
640 REM *** CORRECTION VECTOR, G2 ***
650 FOR I=1 TO F1:S=0:FOR J=1 TO F1
660 S=S+C5(I,J)*G1(J):NEXT J:G2(I)=S:NEXT I:A6,I6=1
670 IF F2=0 THEN 690:PRINT "U1=",U1
680 K6=R:R=0:I6=0:K7=1:A6=.5:GOTO 700
690 PRINT "REFINED PARAMETERS:":PRINT
700 FOR K1=1 TO F1:J=S1(K1):P(J)=P(J)+G2(K1)*A6:NEXT K1
710 GOSUB 1470:GOSUB 830
720 IF I6>0 THEN 800:IF K7=0 THEN 740:U2=U:PRINT "U2=",U2
730 K7=0:GOTO 700
740 U3=U:PRINT "U3=",U3:R=K6:U4=U1-2*U2+U3:IF U4>0 THEN 770
750 A6=1:IF U3<U1 THEN 760:A6=0
760 PRINT "CONCAVE ALFA, IS MADE EQUAL WITH",A6:GOTO 780
770 U4=(U1-U3)/(4*U4):A6=.5+U4:PRINT "ALFA=",A6
780 A6=1-A6:I6=1:IF ABS(A6)<3 THEN 790:A6=2
790 A6=-A6:GOTO 690
800 GOSUB 1280:U1=U:FOR I=1 TO F1:J=S1(I):U7=SQR(ABS(C5(I,I)))*H
810 PRINT "ERROR OF PARAMETER OF NO ";J;":";U7:NEXT I
820 ON V1 GOTO 60,190,230,260,410:END
830 REM EQUSOLV
840 IF R=0 THEN 850:PRINTUSING 1400,P(1);P(2);P(3);P(4);P(5);P(6):PRINT
850 E1=EXP(Z*P(1)):E2=EXP(Z*P(2)):E3=EXP(Z*P(3)):U,S2=0
860 IF W=1 THEN 870:C2,C3,C4=0:A=L:T1=T1-1:E4=EXP(Z*P(4))
870 FOR I=K TO V STEP N:B1=M(I)-N(I)
880 IF W>1 THEN 910:B=.5*(ABS(B1)+SQR(B1^2+4*E1))
890 IF B1>=0 THEN 900:B=E1/B
900 C1(I)=E2*B:C2(I)=E3*E1/B:E1(I)=P(5)+P(6)*Z1*LOG(B)-C1(I)-C2(I):GOTO 1090
910 B=EXP(Z*(E(I)-P(5))/P(6)):A1=L(I):B1=M(I)-N(I)
920 A2=A1*.0011:B2=ABS(B1*.0011):T2,S1=0:IF T1>=0 THEN 930:A=A(I):B=B(I)
930 T2=T2+1:IF T2>8 THEN 1100
940 C1=E1*A^P1*B^Q1:A3=A+C1*P1:B3=B+C1*Q1-J3/B
950 D1=1+C1*P1^2/A:D2=1+C1*Q1^2/B+J3/B^2:D3=C1*P1*Q1
960 ON N3 GOTO 1030,1010,990,970
970 C4=E4*A^P4*B^Q4:A3=A3+C4*P4:B3=B3+C4*Q4
980 D1=D1+C4*P4^2/A:D2=D2+C4*Q4^2/B:D3=D3+C4*P4*Q4
990 C3=E3*A^P3*B^Q3:A3=A3+C3*P3:B3=B3+C3*Q3
```

Appendix C: MINIPOT

```
1000 D1=D1+C3*P3^2/A:D2=D2+C3*Q3^2/B:D3=D3+C3*P3*Q3
1010 C2=E2*A^P2*B^Q2:A3=A3+C2*P2:B3=B3+C2*Q2
1020 D1=D1+C2*P2^2/A:D2=D2+C2*Q2^2/B:D3=D3+C2*P2*Q2
1030 Y1=A3-A1:Y2=B3-B1:D4=D3/A:D3=D3/B:D5=D1*D2-D3*D4:IF D5=0 THEN 1100
1040 A=A+(D3*Y2-D2*Y1)/D5:B=B+(D4*Y1-D1*Y2)/D5
1050 IF ABS(Y1)>A2 THEN 930:IF ABS(Y2)>B2 THEN 930
1060 IF A<0 THEN 1100:IF A>A1 THEN 1100:IF B<0 THEN 1100
1070 A(I)=A:B(I)=B:C1(I)=C1:C2(I)=C2:C3(I)=C3:C4(I)=C4
1080 E1(I)=P(5)+P(6)*Z1*LOG(B)-J1*B-J2*J3/B
1090 D(I)=E(I)-E1(I):S2=S2+1:U=U+D(I)^2:GOTO 1270
1100 A=L:B=EXP(Z*(E(I)-P(5))/P(6)):IF T1>=0 THEN 1110:A=A(I):B=B(I)
1110 G1=2:D5=B
1120 G2=2:D6=A
1130 C1=E1*A^P1*B^Q1:A3=A+C1*P1:B3=B+C1*Q1-J3/B
1140 ON N3 GOTO 1180,1170,1160,1150
1150 C4=E4*A^P4*B^Q4:A3=A3+C4*P4:B3=B3+C4*Q4
1160 C3=E3*A^P3*B^Q3:A3=A3+C3*P3:B3=B3+C3*Q3
1170 C2=E2*A^P2*B^Q2:A3=A3+C2*P2:B3=B3+C2*Q2
1180 IF ABS(A1-A3)<A2 THEN 1210:IF A1-A3<0 THEN 1200
1190 D6=G2*D6:A=A+D6:GOTO 1130
1200 G2=.5:D6=G2*D6:A=A-D6:GOTO 1130
1210 IF B1>0 THEN 1220:B4=B3-B1:GOTO 1230
1220 B4=B1-B3
1230 B5=ABS(B4):IF B2>B5 THEN 1070:IF B1<0 THEN 1260:IF B4<0 THEN 1250
1240 D5=G1*D5:B=B+D5:GOTO 1120
1250 G1=.5:D5=G1*D5:B=B-D5:GOTO 1120
1260 IF B4<0 THEN 1240:GOTO 1250
1270 NEXT I:RETURN
1280 REM PRINT
1290 IF R<=1 THEN 1420:FOR I=K TO V STEP N:IF W>1 THEN 1330
1300 PRINTUSING 1350,I,V1(I),M(I)*E5,N(I)*E5,E(I),E1(I),D(I),C1(I),C2(I)
1310 GOTO 1410
1320 IF R<=1 THEN 1410
1330 PRINTUSING 1340,I,V1(I),M(I)*E5,N(I)*E5,L(I)*E5,E(I),E1(I),D(I)
1340 %####-###.##-###.####-###.####   -###.####-#####.##-#####.##-#####.##
1350 %####-###.##-###.####-###.####-#####.##-#####.##-#####.##-####.##
```

```
1360 IF R<=2 THEN 1410:IF N4>1 THEN 1370:X6=100/M(I):GOTO 1380
1370 X6=100/L(I)
1380 PRINTUSING 1390,C1(I)*X6;C2(I)*X6;C3(I)*X6;C4(I)*X6;100*B(I)/M(I);
     A(I)*100/L(I)
1390 %-###############.##-#####.##-#####.##-#####.##-#####.##-####
1400 %-###############.####-###.####-###.####-###.####-#####.##-####
1410 NEXT I
1420 PRINTUSING 1430,":";U
1430 %SQUARE OF RESIDUALS#   #.####^^^^
1440 H=SQR(U/(S2-F1-1)):PRINT :PRINTUSING 1450,":";H
1450 %STANDARD DEVIATION #   #.####^^^^
1460 PRINT HEX(0A0A0A):RETURN
1470 REM LIGOPT:IF W=2 THEN 1490:IF T6<>4 THEN 1490:E4=EXP(Z*P(4))
1480 FOR I=1 TO N1:N(I)=(V0*M2+V1(I)*E4)/(V0+V1(I)):NEXT I
1490 RETURN
```

Appendix C: MINIPOT

Numerical examples

Two numerical examples of the MINIPOT approach are reproduced by kind permission of Pergamon Press Ltd.

A strong acid (10.0 ml; 0.05 M) was titrated potentiometrically with 0.093626 M strong base in a 1:1 propylene glycol–water mixture. Forty experimental points (titrant volume and glass electrode potential values) were taken (N1 = 40). The values of the following six parameters (N2 = 6) are to be found: log K_s, log j_H, log j_{OH}, log T_{OH}, $E°$ and g.

From the middle part of the titration curve a value of about -60 mV can be calculated for $E°$. The ionic product of water can be taken as the starting value for K_s, 59.15 mV for g, and the analytically determined value for T_{OH}. Small numbers can initially be assigned to j_H and j_{OH}. First, log K_s and $E°$ must be calculated from the middle part of the titration curve (only points 14–33 are used).

Input:

Line 60: 1, 40, 6, 1, 10.0, 0.05, 0, 0, 0, 0.093626, 0,
Line 120:
 0.4, −144.9, 0.6, −146.1, 0.8, −146.7, 1.0, −148.2, 1.2, −148.7, 1.4, −153.2, 1.6, −153.8, 1.8, −154.3, 2.0, −157.5, 2.2, −158.5, 2.4, −161.0, 2.6, −161.5, 2.8, −166.4, 3.0, −167.2, 3.2, −169.3, 3.4, −170.1, 3.6, −174.1, 3.8, −179.9, 4.0, −183.2, 4.2, −189.2, 4.4, −192.8, 4.6, −199.6, 5.0, −224.9, 5.1, −232.8, 5.2, −245.9, 5.3, −271.2, 5.8, −715.6, 5.9, −722.6, 6.0, −728.2, 6.2, −739.3, 6.4, −745.8, 6.6, −750.5, 6.8, −754.2, 7.0, −756.8, 7.2, −760.3, 7.8, −766.8, 8.0, −769.0, 8.4, −772.1, 8.8, −775.0 9.2, −778.0,
Line 210:
 −14, 0.1, 1, 0.03, 1, 0.03, −1.0286, 0.02, −60, 0.1, 59.16, 0.1,
Line 230:
 14, 33, 1,

Task:

Search for log K, (T6 = 1) with (T4 =) 0.1 step (F = 1) between limits (T3 =) −14.2 and (T5 =) −12. After reaching a minimum value for U, execute two further steps (T7 = 2), and then back for new task (V1 = 4).

Input:

Line 260: 1,1,
Line 290: −14.2, 0.1, −12, 1, 1, 2, 4,

Output:

Optimum values: log K_s = −13.7; U = 237.7.

Task:

Search for $E°$ (T6 = 5) with (T4 =) 2 mV step (F = 1) between limits (T3 =) −70 and (T5 =) −50 mV. One step after U_{min} (T7 = 1), and back for new task (V1 = 4).

Input:

Line 260: 1, 3,
Line 290: −70, 2, −50, 5, 1, 1, 4.

Output:

Optimum values: $E°$ = −62; U = 192.3.

Task:

Search for better $E°$ with step and limits determined by its actual value (F = 2). With T3 = 1.05 and T5 = 0.95, the limits will be −65.1 and −58.9, and with T4 = −0.003, the step will be 0.186. (Note that since $E°$ is negative, a negative value must assigned to T4 to obtain a positive value for step.) Back for new task (V1 = 4).

Input:

Line 260: 2, 1,
Line 290: 1.05, −0.003, 0.95, 5, 1, 1, 4.

Output:

Optimum values: $E°$ = −62.68; U = 184.

Task:

Refinement (F = 0) of log K_s [S1(1) = 1] and $E°$ [S1(2) = 5], i.e., of two parameters (F1 = 2) with the calculation of the correction scalar (F2 = 1). New refinement follows (V1 = 5).

Input:
Line 260: 0, 3,
Line 410: 2, 1, 5, 3,
Line 430: 1, 5,

Output:
Optimum values: log $K_s = -13.713$, $E° = -62.46$; U = 181.5.

Task:
The same refinement again, but subsequently the remaining experimental points are included in the calculations (V1 = 3).

Input:
Line 410: 2, 1, 3, 3,
Line 430: 1, 5,

Output:
Optimum values: log $K_s = -13.712$, $E° = -62.46$; U = 181.4.

Input:
Line 230: 1, 40, 1,

Task:
Search for log j_H (T6 = 2) with (T4 =) 0.2 step (F = 1) between limits (T3 =) 1.0 and (T5 =) 5.0. One step after U_{min}(T7 = 1), and back for new task (V1 = 4).

Input:
Line 260: 1, 1,
Line 290: 1.0, 0.2, 5.0, 2, 1, 1, 4,

Output:
Optimum values: log $j_H = 1.4$; U = 374.6.

Task:
The same as the previous one, but the search will be made for log j_{OH} (T6 = 3).

Input:
Line 260: 1, 1,
Line 290: 1.0, 0.2, 5.0, 3, 1, 1, 4,

Output:
Optimum values: log $j_{OH} = 2.6$; U = 194.3.

Task:
Refinement (F = 0) of the four parameters (F1 = 4) log K_s, log j_H, log j_{OH} and $E°$ (the elements of S1 will be 1, 2, 3 and 5), and calculation of the correction scalar (F2 = 1). New refinement follows (V1 = 5).

Input:
Line 260: 0, 1,
Line 410: 4, 1, 5, 3,
Line 430: 1, 2, 3, 5,

Output:
Optimum values: log $K_s = -13.659$, log $j_H = 1.523$, log $j_{OH} = 2.779$. $E° = -62.42$; U = 157.5.

Task:
The same as the previous one, but without optimizing the scalar calculation (F2 = 0). Back for new task (V1 = 4).

Input:
Line 410: 4, 0, 4, 3,
Line 430: 1, 2, 3, 5,

Output:
The same values as the former ones.

Task:
Search for log T_{OH} (T6 = 4) with (T4 =) 0.004 step (F1 = 1) between limits (T3 =) -1.032 and (T5 =) -1.0. One step after U_{min} and back for new task (V1 = 4).

Input:
Line 260: 1, 1,
Line 290: -1.032, 0.004, -1.0, 4, 1, 1, 4,

Output:
Optimum values: log $T_{OH} = -1.0288$; U - 150.6.

Task:
Refinement (F = 0) with optimizing of scalar calculation (F2 = 1) of the four parameters (F1 = 4) log K_s, log j_H, log j_{OH} and $E°$ (S1: 1, 2, 3 and 5), and end of calculation (V1 = 6).

Appendix C: MINIPOT

Input:
Line 260: 0, 1,
Line 410: 4, 1, 6, 3,
Line 430: 1, 2, 3, 5.

Output:
Optimum values: $\log K_s = -13.66 \pm 0.02$, $\log j_H = 1.47 \pm 0.38$, $\log j_{OH} = 2.78 \pm 0.07$, $E° = -62.54 \pm 0.7$; $U = 150.34$, $\sigma = 2.07$ mV.

Problem 2 (W = 2)

A 1.5951×10^{-3} M polypeptide solution was prepared by dissolution in 0.05 M strong acid, and 10.0 ml were titrated with 0.09358 M strong base.

The 17 experimental points used for calculation come from the middle part of the titration curve. In this region four dominant complexes (H_qL_p) are formed, with $(q, p) = (2, 1)$, $(1, 1)$, $(-1, 1)$ and $(-2, 1)$.

Approximate values for free hydrogen-ion and ligand (peptide) concentrations of the individual experimental points are not given (all the elements of arrays B and A are zero, and +1 is assigned to T1). As a starting value for the free peptide concentration ($[L]$), 0.0005 is given. The values of K_s, j_H, j_{OH}, $E°$ and g are those for the input blank curve (problem 1).

Input:
Line 60: 2,
Line 70: 17, 6, 4, 2, 3, 10.0, 0.05, 0, 0.0015951, 0, 0.09358, 0, 0.0005, 29.5, 616.5, 2.1878E-14.

Line 130:
 5.1, −313.0, 0, 0, 5.15, −330.0, 0, 0,
 5.2, −345.3, 0, 0, 5.25, −370.0, 0, 0,
 5.3, −402.0, 0, 0, 5.35, −484.0, 0, 0,
 5.4, −550.0, 0, 0, 5.45, −577.0, 0, 0,
 5.5, −612.0, 0, 0, 5.55, −635.0, 0, 0,
 5.6, −653.0, 0, 0, 5.65, −664.0, 0, 0,
 5.7, −672.4, 0, 0, 5.75, −681.0, 0, 0,
 5.8, −689.7, 0, 0, 5.85, −696.0, 0, 0,
 5.9, −701.2, 0, 0,
Line 190: 2, 1, 1, 1, −1, 1, −2, 1,
Line 210: 7, 0.04, 3, 0.03, −11, 0.04, −21, 0.05, −62.54, 0.1, 59.16, 0.1,
Line 230: 1, 17, 1, 1,

Task:
Search for the approximate $\log \beta$ values with (T4 =) 0.5 step (F = 1) for each, between the limits 7–12, 3–8, −11––5 and −21––15 in order of input. One step after U_{min}. Back for new task in all cases (V1 = 4)

Input:
Line 260: 1, 3,
Line 290: 7, 0.5, 12, 1, 1, 1, 4,
Line 260: 1, 3,
Line 290: 3, 0.5, 8, 2, 1, 1, 4,
Line 260: 1, 3,
Line 290: −11, 0.5, −5, 3, 1, 1, 4,
Line 260: 1, 3,
Line 290: −21, 0.5, −15, 4, 1, 1, 4,

Output:
Optimum $\log \beta$ values: 9.5, 5.0, −8.5 and −18.5; $U = 780$.

Task:
Refinement (F = 0) of the four $\log \beta$ values (F1 = 4, S1: 1, 2, 3 and 4) with optimization of scalar calculation (F2 = 1). Back for new refinement (V1 = 5).

Input:
Line 260: 0, 1,
Line 410: 4, 1, 5, 3,
Line 430: 1, 2, 3, 4,

Output:
Optimum $\log \beta$ values: 9.29, 5.17, −8.47 and −18.7, $U = 319$.

Task:
Repetition of the previous refinement, no optimizing scalar calculation (F2 = 0), and end of calculation (V1 = 6).

Input:
Line 410: 4, 0, 6, 3,
Line 430: 1, 2, 3, 4,

Output:
Final values: $\log \beta_2 = 9.25 \pm 0.09$, $\log \beta_1 = 5.14 \pm 0.07$, $\log \beta_{-1} = -8.47 \pm 0.05$ and $\log \beta_{-2} = -18.73 \pm 0.09$; $U = 318.8$, $\sigma = 5.5$ mV.

Appendix D. The LIGEZ program (R. G. Torrington, unpublished information)

Theory

We give the general equation for the protonation of a ligand L which contains NDP displaceable protons and whose totally protonated state is $H_{NS}L$ where NDP $\leq$ NS.

In this derivation, all charges have been ignored for simplicity of writing and β_{101}, $\beta_{102}, \ldots, \beta_{p0q}$ will be written as $\beta_1, \beta_2, \ldots, \beta_q$.

Consider a solution made by adding V_t cm^3 of strong base of concentration [SB] mol dm^{-3} to V_{init} cm^3 of a mixture which has a concentration of [MA] mol dm^{-3} towards a strong monoprotic acid and [TL] mol dm^{-3} towards the species $H_{NDP}L$, the following mass balance relationships apply:

$$T_H = \frac{[MA] \times V_{init}}{V_{init} + V_t} + \frac{NDP \times [TL] \times V_{init}}{V_{init} + V_t} - \frac{V_t[SB]}{V_{init} + V_t}$$

$$= [H] - [OH] + [HL] + 2[H_2L] + \cdots NDP[H_{NDP}L] + \cdots NS[H_{NS}L]$$

$$T_L = \frac{[TL]V_{init}}{V_{init} + V_t} = [L] + [HL] + \cdots [H_{NS}L]$$

where [X] represents the concentrations of the species X in mol dm^{-3}.

The following ligand proton equilibria are also established:

$$H + L = HL \quad \text{for which } \beta_1 = \frac{[HL]}{[H][L]}$$

$$2H + L = H_2L \quad \text{for which } \beta_2 = \frac{[H_2L]}{[H]^2[L]}$$

$$\vdots$$

$$NSH + L = H_{NS}L \quad \text{for which } \beta_{NS} = \frac{[H_{NS}L]}{[H]^{NS}[L]}.$$

Also

$$K_W = [H][OH]$$

$$T_L = [L] + \beta_1[H][L] + \beta_2[H]^2[L] + \cdots + \beta_{NS}[H]^{NS}[L]$$

$$= [L][1 + \beta_1[H] + \beta_2[H]^2 + \cdots + \beta_{NS}[H]^{NS}]$$

$$T_H = [H] - [OH] + \beta_1[H][L] + 2\beta_2[H]^2[L] + \cdots + NS\beta_{NS}[H]^{NS}[L]$$

$$T_H - [H] + [OH] = [L][\beta_1[H] + 2\beta_2[H]^2 + \cdots + NS\beta_{NS}[H]^{NS}]$$

$$[L] = [T_H - [H] + [OH]]/[\beta_1[H] + 2\beta_2[H]^2 + \cdots + NS\beta_{NS}[H]^{NS}]$$

Appendix D: LIGEZ

$$T_L = [T_H - [H] + [OH]]/[\beta_1[H] + \cdots + NS\beta_{NS}[H]^{NS}][1 + \beta_1[H] + \cdots + \beta_{NS}[H]^{NS}]$$

$$T_L\beta_1[H] + \cdots + T_L NS\beta_{NS}[H]^{NS} = T_H - [H] + [OH] + \beta_1[H]T_H - \beta_1[H]^2 + \beta_1[H][OH]$$
$$+ \cdots + T_H\beta_{NS}[H]^{NS} - \beta_{NS}[H]^{NS+1} + \beta_{NS}[H]^{NS}[OH]$$

$$T_L\beta_1[H] + \cdots + T_L NS\beta_{NS}[H]^{NS} = T_H - [H] + \frac{K_W}{[H]} + \beta_1[H]T_H - \beta_1[H]^2 + \beta_1 K_W$$
$$+ \cdots + T_H\beta_{NS}[H]^{NS} - \beta_{NS}[H]^{NS+1} + \beta_{NS}K_W[H]^{NS-1}.$$

Multiplying both sides by [H] gives:

$$T_L\beta_1[H]^2 + \cdots + T_L NS\beta_{NS}[H]^{NS+1} = T_H[H] - [H]^2 + K_W + \beta_1[H]^2 T_H$$
$$- \beta_1[H]^3 + \beta_1 K_W[H] + \cdots + T_H\beta_{NS}[H]^{NS+1} - \beta_{NS}[H]^{NS+2} + \beta_{NS}K_W[H]^{NS}.$$

Rearrangement gives:

$$K_W + [T_H + \beta_1 K_W][H] + [T_H\beta_1 - 1 + \beta_2 K_W - T_L\beta_1][H]^2$$
$$+ \cdots + [T_H\beta_{NS-1} - \beta_{NS-2} + \beta_{NS}K_W - [NS-1]T_L\beta_{NS-1}][H]^{NS}$$
$$+ [T_H\beta_{NS} - \beta_{NS-1} NS\, T_L\beta_{NS}][H]^{NS+1} - \beta_{NS}[H]^{NS+2}.$$

This polynomial in [H] can be summarized as

$$K_W + \sum_{n=0}^{NS+1} (T_H\beta_n - \beta_{n-1} + \beta_{n+1}K_W - nT_L\beta_n)[H]^{n+1} = 0 \qquad (D.1)$$

where $\beta_{-1} = \beta_{NS+1} = \beta_{NS+2} = 0$ and $\beta_0 = 1$.

If $[MA]_{init}$, $[TL]_{init}$, $[SB]$, K_W and the β values are known, this equation can be solved for [H] by using numerical techniques such as Newton's method.

When the system under consideration is the titration of a strong acid with a strong base, then NS = NDP = 1 and $\beta_1 = 0$ because the substance, a strong acid, is completely ionized; also $T_L = 0$. Equation (D.1) then becomes

$$K_W + (T_H\beta_0 - \beta_{-1} + \beta_1 K_W)[H] + (T_H\beta_1 - \beta_0 + \beta_2 K_W - T_L\beta_1)[H]^2$$
$$+ (T_H\beta_2 - \beta_1 + \beta_3 K_W - 2T_L\beta_2)[H]^3 = 0.$$

Substitution of $\beta_0 = 1$, $\beta_{-1} = \beta_1 = \beta_2 = \beta_3 = 0$, $T_L = 0$, gives

$$K_W + T_H[H] - [H]^2 = 0.$$

For a monoprotic weak acid such as acetic for which NS = NDP = 1, $\beta_1 = \beta \neq 0$, we obtain

$$K_W + (T_H + \beta K_W)[H] + (T_H\beta - 1 - T_L\beta)[H]^2 - \beta[H]^3 = 0.$$

A diprotic acid such as glycine, for which $\beta_1 \neq 0$, $\beta_2 \neq 0$, NS = 2, and NDP = 1, gives

$$K_W + (T_H + \beta_1 K_W)[H] + (T_H\beta_1 - 1 + \beta_2 K_W - T_L\beta_1)[H]^2 + (T_H\beta_2 - \beta_1 - 2T_L\beta_2)[H]^3$$
$$- \beta_2[H]^4 = 0.$$

If in the equation

$$E = E_{\text{const}} + s \ln [H]$$

E is measurable, E_{const} and s are unknown but constant, and [H] is a variable, we can write

$$E = (E'_{\text{const}} + \Delta E_{\text{const}}) + (s' + \Delta s) \ln ([H]' + \Delta[H])$$

where the primed values are estimates of E_{const}, s and [H] and ΔE_{const}, Δs and ΔH are the differences between the true values and the estimated values.
Then

$$E = E'_{\text{const}} + \Delta E_{\text{const}} + s' \ln ([H]' + \Delta[H]) + \Delta s \ln ([H]' + \Delta[H])$$

$$= E'_{\text{const}} + \Delta E_{\text{const}} + s' \ln [H]' \left(1 + \frac{\Delta[H]}{[H]'}\right) + \Delta s \ln [H]' \left(1 + \frac{\Delta[H]}{[H]'}\right)$$

$$= E'_{\text{const}} + \Delta E_{\text{const}} + s' \ln [H]' + s' \frac{\Delta[H]}{[H]'} + \Delta s \ln [H]'$$

if it can be assumed that $\Delta[H]/[H]' \ll 1$ such that

$$\ln \left(1 + \frac{\Delta[H]}{[H]'}\right) \simeq \frac{\Delta[H]}{[H]'} \quad \text{and} \quad \frac{\Delta s [\Delta H]}{[H]'} = 0$$

$$\Delta E = E - E'_{\text{const}} - s' \ln [H]' = \Delta E_{\text{const}} + \Delta s \ln [H]' + \frac{s'}{[H]'} \Delta[H], \qquad (D.2)$$

Equation (D.1) can be written

$$f = K_W + \sum_{n=0}^{NS+1} (T_H \beta_n - \beta_{n-1} + \beta_{n+1} K_W - n T_L \beta_n)[H]^{n+1} = 0$$

$$df = 0 = \frac{\delta f}{\delta[H]} d[H] + \frac{\delta f}{\delta \beta_1} d\beta_1 + \cdots \frac{\delta f}{\delta \beta_{NS}} d\beta_{NS} + \frac{\delta f}{\delta[MA]} d[MA] + \frac{\delta f}{\delta[TL]} d[TL]$$

$$+ \frac{\delta f}{\delta[SB]} d[SB] + \frac{\delta f}{\delta K_W} dK_W.$$

Hence,

$$d[H] = -\left(\frac{\delta f}{\delta \beta_1} d\beta_1 + \cdots + \frac{\delta f}{\delta K_W} dK_W\right) \bigg/ \frac{\delta f}{\delta[H]}.$$

If the differentials can now be replaced by deltas, we obtain

$$\Delta[H] = -\left(\frac{\delta f}{\delta \beta_1} \Delta\beta_1 + \cdots + \frac{\delta f}{\delta K_W} \Delta K_W\right) \bigg/ \frac{\delta f}{\delta[H]}$$

Appendix D: LIGEZ

where the partial differentials are to be enumerated at the estimated values of the parameters.

Thus, equation (D.2) becomes:

$$\Delta E = \Delta E_{const} + \Delta s \ln [H]' + \frac{s'}{[H]'}\left[-\left(\frac{\delta f}{\delta \beta_1}\Delta \beta_1 + \cdots + \frac{\delta f}{\delta K_W}\Delta K_W\right)\Big/\frac{\delta f}{\delta [H]}\right]$$

$$= \Delta E_{const} + \Delta s \ln [H]' + \frac{s'}{[I I]'}\left(-\frac{\delta f}{\delta \beta_1}\Big/\frac{\delta f}{\delta [H]}\right)\Delta \beta_1$$

$$+ \cdots + \left(-\frac{s'}{[H]'}\frac{\delta f}{\delta K_W}\Big/\frac{\delta f}{\delta [H]}\right)\Delta K_W.$$

This is of the form

$$\Delta E = \Delta E_{const} + A \Delta s + B \Delta \beta_1 + C \Delta \beta_2 + \cdots + I \Delta K_W$$

where $A, B, C, \ldots, I$ are variables which can be calculated at each point and the Δs are corrections to the estimated values of the parameters and can now be found by the normal linear regression method.

The LIGEZ program coding

There follows the coding and the method for running the program LIGEZ on a HP41C calculator fitted with a Quadram memory module and a Maths Pac Module.

The program assumes that the concentration of hydrogen can be calculated from the equation

$$K_W + (T_H + \beta_{101} K_W)[H] + (T_H \beta_{101} - 1 + \beta_{102} K_W - T_L \beta_{101})[H]^2$$
$$+ (T_H \beta_{102} - \beta_{101} - 2 T_L \beta_{102})[H]^3 - \beta_{102}[H]^4 = 0$$

and that the electrode equation is of the form

$$E = E_{const} - s \ln (1 + b[H]) + s \ln (a + [H])$$

when b and a are both 0, this reduces to

$$E = E_{const} + s \ln [H]$$

which is the equation that was used in the examples given in the text.

LIGEZ Coding for HP41C Calculator

01	LBL LIGEZ	07	STO 95	13	XEQ 30	19	×
02	RCL 97	08	0	14	RCL 78	20	RCL 90
03	STO 00	09	STO 96	15	RCL 90	21	×
04	RCL 92	10	XEQ SET	16	×	22	+
05	STO 93	11	XEQ CLEAR	17	RCL 91	23	RCL 91
06	RCL 94	12	LBL 13	18	RCL 86	24	RCL 79

#	Op	#	Op	#	Op	#	Op
25	×	76	×	127	RCL 06	178	RCL 00
26	RCL 09	77	RCL 84	128	RCL 10	179	RCL 87
27	×	78	−	129	÷	180	+
28	+	79	2	130	−	181	÷
29	RCL 88	80	RCL 02	131	STO 11	182	LN
30	RCL 09	81	×	132	RCL 00	183	CHS
31	×	82	RCL 85	133	−	184	STO IND 01
32	−	83	×	134	ABS	185	1
33	RCL 90	84	−	135	RCL 00	186	STO+01
34	RCL 09	85	STO 05	136	÷	187	LBL 29
35	+	86	LBL 09	137	STO 12	188	FS? 18
36	÷	87	RCL 00	138	RCL 11	189	GTO 14
37	STO 08	88	ENTER	139	STO 00	190	GTO 15
38	RCL 86	89	ENTER	140	1E−05	191	LBL 14
39	RCL 90	90	ENTER	141	RCL 12	192	RCL 00
40	×	91	RCL 85	142	x≤y?	193	x^2
41	RCL 79	92	×	143	GTO 08	194	RCL 85
42	RCL 09	93	RCL 05	144	GTO 09	195	×
43	×	94	−	145	LBL 08	196	RCL 00
44	+	95	×	146	RCL 99	197	RCL 84
45	RCL 90	96	RCL 04	147	RCL 87	198	×
46	RCL 09	97	−	148	×	199	+
47	+	98	×	149	RCL 83	200	1
48	÷	99	RCL 03	150	×	201	+
49	STO 02	100	−	151	RCL 99	202	RCL 10
50	RCL 79	101	×	152	−	203	×
51	RCL 86	102	RCL 82	153	RCL 00	204	STO IND 01
52	÷	103	−	154	RCL 83	205	1
53	STO 13	104	STO 06	155	×	206	STO+01
54	RCL 84	105	RCL 00	156	1	207	LBL 15
55	RCL 82	106	ENTER	157	+	208	FS?12
56	×	107	ENTER	158	RCL 00	209	GTO 16
57	RCL 08	108	ENTER	159	RCL 87	210	GTO 17
58	+	109	4	160	+	211	LBL 16
59	STO 03	110	RCL 85	161	×	212	RCL 99
60	RCL 08	111	×	162	÷	213	RCL 00
61	RCL 84	112	×	163	RCL 10	214	×
62	×	113	RCL 05	164	÷	215	RCL 83
63	1	114	3	165	CHS	216	RCL 00
64	−	115	×	166	STO 10	217	×
65	RCL 85	116	−	167	RCL 80	218	1
66	RCL 82	117	×	168	STO 01	219	+
67	×	118	RCL 04	169	FS?11	220	÷
68	+	119	2	170	GTO 28	221	CHS
69	RCL 02	120	×	171	GTO 29	222	STO IND 01
70	RCL 84	121	−	172	LBL 28	223	1
71	×	122	×	173	RCL 00	224	STO+01
72	−	123	RCL 03	174	RCL 83	225	LBL 17
73	STO 04	124	−	175	×	226	FS?13
74	RCL 08	125	STO 10	176	1	227	GTO 18
75	RCL 85	126	RCL 00	177	+	228	GTO 19

Appendix D: LIGEZ

229	LBL 18	280	×	331	LBL 25	382	100
230	RCL 99	281	STO IND 01	332	FS?17	383	×
231	RCL 00	282	1	333	GTO 26	384	ABS
232	RCL 87	283	STO+01	334	GTO 27	385	STO 09
233	+	284	LBL 23	335	LBL 26	386	END
234	÷	285	FS?16	336	1	01	LBL SET
235	STO IND 01	286	GTO 24	337	STO IND 01	02	0
236	1	287	GTO 25	338	LBL 27	03	STO 14
237	STO+01	288	LBL 24	339	XEQ 30	04	11
238	LBL 19	289	RCL 00	340	RCL 10	05	STO 02
239	RCL 00	290	3	341	RCL 89	06	LBL 10
240	ENTER	291	y^x	342	−	07	RCL 02
241	ENTER	292	RCL 85	343	RCL 00	08	19
242	ENTER	293	×	344	RCL 83	09	x=y?
243	1	294	RCL 00	345	×	10	GTO 12
244	CHS	295	x^2	346	1	11	FS?IND 02
245	×	296	RCL 84	347	+	12	GTO 11
246	RCL 08	297	×	348	LN	13	1
247	RCL 02	298	+	349	RCL 99	14	STO+02
248	−	299	RCL 00	350	×	15	GTO 10
249	+	300	+	351	+	16	LBL 11
250	×	301	RCL 91	352	RCL 00	17	1
251	RCL 82	302	×	353	FS?06	18	STO+02
252	+	303	RCL 00	354	STOP	19	STO+14
253	×	304	3	355	RCL 87	20	GTO 10
254	STO 03	305	y^x	356	+	21	LBL 12
255	FS?14	306	2	357	LN	22	RCL 14
256	GTO 20	307	×	358	RCL 99	23	x^2
257	GTO 21	308	RCL 85	359	×	24	STO 11
258	LBL 20	309	×	360	−	25	15
259	RCL 03	310	RCL 00	361	STO 08	26	+
260	RCL 10	311	x^2	362	FS?19	27	STO 80
261	×	312	RCL 84	363	STOP	28	RCL 14
262	STO IND 01	313	×	364	x^2	29	+
263	1	314	+	365	STO+96	30	STO 81
264	STO+01	315	−	366	XEQ SUM	31	RTN
265	LBL 21	316	RCL 13	367	RCL 98	32	LBL CLEAR
266	FS? 15	317	RCL 09	368	STO+95	33	RCL 11
267	GTO 22	318	×	369	RCL 93	34	RCL 14
268	GTO 23	319	RCL 90	370	PSE	35	2
269	LBL 22	320	+	371	DSE 93	36	×
270	RCL 03	321	RCL 90	372	GTO 13	37	+
271	RCL 00	322	RCL 09	373	RTN	38	1000
272	×	323	+	374	LBL 30	39	÷
273	RCL 00	324	÷	375	RCL IND 95	40	1.00001
274	3	325	×	376	INT	41	+
275	y^x	326	RCL 10	377	10	42	STO 01
276	RCL 02	327	×	378	÷	43	15
277	×	328	STO IND 01	379	STO 10	44	STO 02
278	−	329	1	380	RCL IND 95	45	LBL 01
279	RCL 10	330	STO+01	381	FRC	46	0

47	STO IND 02	73	1	99	1	125	STO+12
48	1	74	STO-07	100	RCL 07	126	STO+02
49	STO+02	75	RCL 07	101	x=y?	127	DSE 13
50	ISG 01	76	.00001	102	GTO 05	128	GTO 04
51	GTO 01	77	+	103	.00101	129	RCL 81
52	RTN	78	STO 06	104	+	130	STO 01
53	LBL SUM	79	LBL 02	105	STO 06	131	RCL 14
54	0	80	RCL IND 02	106	LBL 03	132	.00001
55	STO 09	81	RCL IND 03	107	RCL IND 11	133	+
56	1	82	×	108	STO IND 01	134	STO 06
57	STO 12	83	STO+IND 01	109	1	135	RCL 80
58	RCL 14	84	1	110	STO+11	136	STO 02
59	1	85	STO+01	111	RCL 14	137	LBL 06
60	+	86	STO+03	112	STO+01	138	RCL 08
61	STO 07	87	DSE 06	113	DSE 06	139	RCL IND 02
62	RCL 80	88	GTO 02	114	GTO 03	140	×
63	STO 02	89	RCL 09	115	LBL 05	141	STO+IND 01
64	15	90	STO+01	116	RCL 14	142	1
65	STO 01	91	1	117	1	143	STO+02
66	RCL 14	92	STO+09	118	+	144	STO+01
67	0.00001	93	RCL 01	119	RCL 12	145	DSE 06
68	+	94	RCL 14	120	×	146	GTO 06
69	STO 13	95	−	121	15	147	END
70	LBL 04	96	1	122	+		
71	RCL 02	97	+	123	STO 01		
72	STO 03	98	STO 11	124	1		

Running the LIGEZ program

The program uses approximately 100 storage registers in addition to those registers used for storing the data being processed. Hence, if 60 points are to be handled, a suitable *size* would be 160.

The data, therefore, are placed in registers 100→159. The data must be put into the following format before being placed in the data registers; assuming that the emf value is $xyz \cdot w$, and that the volume value is $ab \cdot cd$, the point is then written as $xyzw \cdot abcd$, this value being placed in the data register.

INPUT

(a) *Parameters not being refined*

    (i) 1.0000, store register 98
    (ii) Initial volume, store register 90
    (iii) NDP, store register 91
    (iv) Starting estimate of hydrogen ion concentration, store register 97
    (v) Concentration of strong acid when this is added to a weak acid system, store register 78
    (vi) Concentration of ligand in burette, store 79
    (vii) Number of datapoints to be processed, store as $nn \cdot 00001$ in register 92
    (viii) Address of first data point, store in register 94
    (ix) Concentration of alkali in burette, store register 88

Appendix D: LIGEZ

(b) *Parameters which can be refined*

The following parameters will be refined if the flag indicated is set. The parameters are given in the order in which the corrections to the estimated values will appear when the simultaneous equations are solved.

(i) s, store in register 99 and set flag 11 for refinement
(ii) K_W, the ionic product of water, store in register 82 and set flag 18 for refinement
(iii) b of the electrode equation above, store in register 83 and set flag 12 for refinement
(iv) a of the electrode equation above, store in register 87 and set flag 13 for refinement
(v) β_{101}, store in register 84 and set flag 14 for refinement
(vi) β_{102}, store in register 85 and set flag 15 for refinement
(vii) Concentration of ligand, store in register 86 and set flag 16 for refinement
(viii) E_{const}, store in register 89 and set flag 17 for refinement

Note. A maximum of 7 of the 8 parameters above can be refined at one time if the calculator program is being used.

Program execution

(1) Enter all parameters and set the refinement flags
(2) Enter the data to be processed
(3) Press XEQ ALPHA LIGEZ ALPHA

The program will now run. The progress of the run can be followed by the numbers which appear in the display after the last point has been processed. When the number 1.0001 appears, the run will stop and the simultaneous equations must now be solved. This is done using the Maths Pac module. After the run stops the following instructions are entered from the keyboard.

1 SF 04
2 SF 05
3 SF 21
4 XEQ ALPHA SIMEQ ALPHA

The calculator then prompts for the values of the column. Press R/S after each prompt.

The calculator then solves the simultaneous equations for the corrections to be applied to the parameters being refined.

Those corrections can then be added to the parameters and the whole procedure restarted for the next cycle.

The value of $\sum (E_{obs} - E_{calc})^2$ is stored after each run in register 96 and can be obtained by recalling that register.

During the processing of a point, the calculated hydrogen ion concentration can be obtained if flag 6 is set. The value of $E_{obs} - E_{calc}$ can also be obtained by setting flag 19. However, the program then halts and must be restarted by pressing R/S.

Table D1. Data for example 1 using LIGEZ. Simulated titration data for a glycine-like ligand titrated with a strong base in the presence of a strong acid. [NaOH] = 19.42 mmol dm^{-3}

Volume base (cm^3)	emf (mV)	Volume base (cm^3)	emf (mV)	Volume base (cm^3)	emf (mV)
0.00	291.2	3.60	256.7	8.00	−123.3
0.20	289.8	3.80	253.4	8.20	−126.9
0.40	288.4	4.00	249.8	8.40	−130.3
0.60	287.0	4.20	245.7	8.60	−133.7
0.80	285.5	4.40	240.9	8.80	−137.1
1.00	284.0	4.60	235.2	9.00	−140.5
1.20	282.4	4.80	228.1	9.20	−143.9
1.40	280.8	5.00	218.7	9.40	−147.4
1.60	279.1	6.00	−71.1	9.60	−150.8
1.80	277.4	6.20	−80.3	9.80	−154.4
2.00	275.6	6.40	−87.6	10.00	−157.9
2.20	273.7	6.60	−93.7	10.20	−161.5
2.40	271.7	6.80	−99.0	10.40	−165.0
2.60	269.6	7.00	−103.8	10.60	−168.6
2.80	267.3	7.20	−108.1	10.80	−172.0
3.00	265.0	7.40	−112.2		
3.20	262.4	7.60	−116.1		
3.40	259.7	7.80	−119.8		

Table D2. Data for example 2 using LIGEZ. Titration of acetic acid with sodium hydroxide. [NaOH] = 19.46 mmol dm^{-3}

Volume base (cm^3)	emf (mV)	Volume base (cm^3)	emf (mV)	Volume base (cm^3)	emf (mV)
0.00	234.4	3.00	182.6	6.00	151.3
0.20	229.5	3.20	180.3	6.20	149.2
0.40	224.8	3.40	178.0	6.40	147.1
0.60	220.2	3.60	175.8	6.60	145.0
0.80	216.0	3.80	173.6	6.80	142.8
1.00	212.0	4.00	171.5	7.00	140.4
1.20	208.2	4.20	169.5	7.20	138.0
1.40	204.7	4.40	167.4	7.40	135.5
1.60	201.4	4.60	165.5	7.60	132.9
1.80	198.3	4.80	163.4	7.80	130.2
2.00	195.3	5.00	161.5	8.00	127.3
2.20	192.6	5.20	159.5	8.20	124.2
2.40	190.0	5.40	157.4	8.40	120.8
2.60	187.4	5.60	155.5	8.60	117.2
2.80	184.9	5.80	153.4	8.80	113.2

References

Ahrland, S. 1975. In *The Nature of Sea Water*, ed. E. D. Goldberg. Berlin, Dahlem Konferenzen
Al-Falahi, H., May, P. M., Roe, A. M., Slater, R. A., Trott, W. J., and Williams, D. R. 1984. *Agents and Actions*, **14,** 113–120.
Allison, S. A., Harris, P. J., and Marsicano, F. 1975. *J. S. A. Inst. Mining and Metallurgy*, 123–4
Almgren T., Dyrssen, D., and Strandberg, M. 1975. *Deep Sea Res.*, **22,** 635–46
Anderegg, G. 1977. *Critical Survey of Stability Constants of EDTA*. Oxford, Pergamon
Antelman, M. S. 1981. *The Encyclopaedia of Chemical Electrode Potentials*. New York, Plenum
Arena, G., Musumeci, S., Rizzarelli, E., Sammartano, S., and Rigano, C. 1979. *Talanta*, **26,** 1–14
Arnek, R., Sillén, L. G., and Wahlberg, O. 1969. *Arkiv. Kemi* **31,** 353–63
Atkins, P. W. 1982. *Physical Chemistry*, 2nd edn. Oxford University Press
Ball, J. W., Jenne, E. A., and Nordstrom, D. K. 1979. *Chemical Modeling in Aqueous Systems*, ed. E. A. Jenne. Amer. Chem. Soc. Symposium Series **93**
Bates, R. G. 1954. *Electrometric pH Determinations*. New York, Wiley
Bates, R. G. 1973. *Determination of pH, Theory and Practice*. New York, Wiley
Bates, R. G. 1981. *Crit. Rev. Anal. Chem.*, **10,** 247–78
Bates, R. G. 1982. *Pure Appl. Chem.*, **54,** 229–32
Bates, R. G. and Calais, J. G. 1981. *J. Solution Chem.*, **10,** 269–79
Bates, R. G. and Guggenheim, E. A. 1960. *Pure Appl. Chem.*, **1,** 163–8
Beck, M. T. 1977. *Pure Appl. Chem.*, **49,** 127–35
Berthon, G., May, P. M., and Williams, D. R. 1978. *J. Chem. Soc. Dalton*, 1433–8
Brauner, P., Sillén, L. G., Warnqvist, B., and Whiteker, R. 1969. *Arkiv. Kemi*, **31,** 365–76
Bütikofer, H. P. and Covington, A. K. 1979. *Anal. Chim. Acta*, **108,** 179–91
Cole, A., May, P. M., and Williams, D. R. 1981. *Agents and Actions*, **11,** 296–305
Conte, S. D. and de Boer, C. 1972. *Elementary Numerical Analysis: An Algorithmic Approach*, 2nd edn. New York, McGraw-Hill
Covington, A. K. 1980. *Chem. Internat.*, Issue No **6,** 23–4

Covington, A. K. 1981. *Anal. Chim. Acta*, **127**, 1–21
Covington, A. K., Bates, R. G. and Durst, R. A. 1983. *Pure Appl. Chem.*, **55**, 1467–76
Covington, A. K., Paabo, M., Robinson, R. A., and Bates, R. G. 1968. *Anal. Chem.*, **40**, 700–706
Crow, D. R. 1979. *Lab. Practice*, **28**(11), 1209–13
Czerminski, J. B., Dickson, A. G. and Bates, R. G. 1982. *J. Solution Chem.*, **11**, 79–89
Dole, M. 1941. *The Glass Electrode*. London, Chapman and Hall
Dyrssen, D. and Wedborg, M. 1974. In *The Sea*, ed. E. D. Goldberg, vol. 5. New York, Wiley
Dyrssen, D., Jagner, D., and Wengelin, F. 1968. *Computer Calculation of Ionic Equilibria and Titration Procedures*. London, Wiley
Eisenman, G. (ed.) 1967. *Glass Electrodes for Hydrogen and Other Cations*. London, Arnold
Florence, T. M. 1982. *Talanta*, **29**, 345–64
Ford, M. A. 1982. As reported in *Chem. Internat.*, Issue No **4**, 5–6
Gaizer, F. and Puskas, A. 1981. *Talanta*, **28**, 565–73
Gans, P. 1976. *Coord. Chem. Rev.*, **19**, 99–124
Gans, P., Sabatini, A., and Vacca, A. 1976. *Inorg. Chim. Acta*, **18**, 237–9
Gans, P., Sabatini, A., and Vacca, A. 1983. *Inorg. Chim. Acta*, **79**, 219–20
Glascoe, P. K. and Long, F. A. 1960. *J. Phys. Chem.*, **64**, 188–90
Glasstone, S. and Lewis, D. 1960. *Elements of Physical Chemistry*. London, Macmillan
Goldberg, E. D. 1963. *The Sea*, ed. E. D. Goldberg, vol. 2, New York, Wiley
Goldschmidt, V. M. 1933. *Fortschr. Min. Krist. Petrogr.*, **17**, 112–56
Gran, G. 1950. *Acta Chem. Scand.*, **4**, 559–73
Gran, G. 1952. *Analyst.*, **77**, 661–9
Guggenheim, E. A. 1930. *J. Phys. Chem.*, **34**, 1758–66
Halstead, B. and Williams, D. R. 1983. *J. Toxicol. Clin. Toxicol.*, **19**, 1081–115
Hamer, W. J. and Acree, S. F. 1944. *J. Res. Natl. Bur. Stand.*, **32**, 215–27
Hamer, W. J., Pinching, G. D., and Acree, S. F. 1945. *J. Res. Natl. Bur. Stand.*, **35**, 539–64
Hamer, W. J., Pinching, G. D., and Acree, S. F. 1946. *J. Res. Natl. Bur. Stand.*, **36**, 47–62
Hamilton, W. C. 1964. *Statistics in Physical Sciences*. New York, Ronald Press
Hansson, I., Ahrland, S., Bates, R. G., Biedermann, G., Dyrssen, D., Högfeldt, E., Martell, A. E., Morgan, J. J., Schindler, P. W., Warner, T. B. and Whitfield, M. 1975. *The Nature of Sea Water*, ed. E. D. Goldberg. Berlin, Dahlem Konferenzen
Harris, P. J. and Marsicano, F. 1975. National Institute for Metallurgy, Report No. **1721**
Hartley, F. R., Burgess, C., and Alcock, R. 1980. *Solution Equilibria*. Chichester, Ellis Horwood
Högfeldt, E. 1982. *Stability Constants of Metal-Ion Complexes. Part A: Inorganic Ligands*. Oxford, Pergamon
Horn, M. K. and Adams, J. A. S. 1966. *Geochim. Cosmochim. Acta*, **30**, 279–97
Huang, Z-X., May, P. M., Quinlan, K. M., Williams, D. R., and Creighton, A. M. 1982. *Agents and Actions*, **12**, 103–109
Hughes, M. A. and Williams, D. R. 1984. (Unpublished data.)
Ingri, N. and Sillén, L. G. 1964. *Arkiv. Kemi*, **23**, 97–121

Ingri, N., Kakolowicz, W., Sillén, L. G., and Warnqvist, B. 1967. *Talanta*, **14**, 1261–86
IUPAC 1979. *Manual of Symbols and Terminology for Physicochemical Quantities and Units*. Oxford, Pergamon
IUPAC. 1980. *Chem. Internat.*, **6**, 23–5
James, A. M. and Pritchard, F. E. 1974. *Practical Physical Chemistry*, 3rd edn. London, Longman
Khoo, K. H., Ramette, R. W., Culberson, C. H. and Bates, R. G. 1977. *Anal. Chem.*, **49**, 29–34
Kielland, J. 1937. *J. Amer. Chem. Soc.*, **59**, 1675–8
Levitt, B. P. 1973. *Findlay's Practical Physical Chemistry*. London, Longman
Linder, P. W. and Murray, K. 1982. *Talanta*, **29**, 377–82
Linder, P. W., May, P. M., Pay, M., Torrington, R. G., and Williams, D. R. 1983. (Unpublished information.)
Lindsay, W. L. 1979. *Chemical Equilibria in Soils*. New York, Wiley
Linnet, N. 1970. *pH Measurements in Theory and Practice*. Copenhagen, Radiometer A/S
McDuff, R. E. and Morel, F. M. M. 1980. *Environ. Sci. Tech.*, **14**, 1182–6
McGlashan, M. L. 1971. *Physico-Chemical Quantities and Units*. Royal Institute of Chemistry Monographs for Teachers **15**
MacInnes, D. A. 1919. *J. Amer. Chem. Soc.*, **41**, 1086–92
Magnusson, V. R., Harris, D. K., Sun, M. S., Taylor, D. K., and Glass, G. E. 1979. In *Chemical Modeling in Aqueous Systems*, ed. E. A. Jenne. Amer. Chem. Soc. Symposium Series **93**
Marcus, Y. 1980. *Rev. Analyt. Chem.*, **4**, 53–137
Marsicano, F., Harris, P. J., McDougall, G. M., and Finkelstein, N. P. 1976. National Institute for Metallurgy, Report No. **1785**
Mattigod, S. V. and Sposito, G. 1979. In *Chemical Modeling in Aqueous Systems*, ed. E. A. Jenne. Amer. Chem. Soc. Symposium Series **93**
Mattock, G. and Taylor, G. R. 1961. *pH Measurement and Titration*. London, Heywood
May, P. M. and Williams, D. R. 1977. *FEBS Lett.*, **78**, 134–8
May, P. M., Linder, P. W., and Williams, D. R. 1977. *J. Chem. Soc. Dalton*, 588–95
May, P. M., Williams, D. R., Linder, P. W., and Torrington, R. G. 1982. *Talanta*, **29**, 249–56
Nancollas, G. H. and Tomson, M. B. 1982. *Pure Appl. Chem.*, **54**, 2675–92
Nelder, J. A. and Mead, R. 1965. *Comput. J.*, **7**, 308–13
Nordstrom, D. K., Jenne, E. A., and Ball, J. W. 1979a. In *Chemical Modeling in Aqueous Systems*, ed. E. A. Jenne. Amer. Chem. Soc. Symposium Series **93**
Nordstrom, D. K., Plummer, L. N., Wigley, T. M. L., Wolery, T. J., Ball, J. W., Jenne, E. A., Bassett, R. L., Crerar, D. A., Florence, T. M., Fritz, B., Hoffman, M., Holdren, G. J. Jr., Lafon, G. M., Mattigod, S. V., McDuff, R. E., Morel, F., Reddy, M. M., Sposito, G., and Thrailkill, J. 1979b. In *Chemical Modeling in Aqueous Systems*, ed. E. A. Jenne. Amer. Chem. Soc. Symposium Series **93**
Paabo, M. and Bates, R. G. 1969. *Anal. Chem.*, **41**, 283–5
Pantony, D. A. 1961. *Statistics, Theory of Error and Design of Experiment*. Royal Institute of Chemistry Lecture Series **2**
Perrin, D. D. 1979. *Stability Constants of Metal-Ion Complexes. Part B: Organic Ligands*. Oxford, Pergamon
Perrin, D. D. and Dempsey, B. 1974. *Buffers for pH and Metal Ion Control*. London, Chapman and Hall

Perrin, D. D. and Sayce, I. G. 1967. *Talanta*, **14**, 833–42
Ramette, R. W., Culberson, C. H. and Bates, R. G. 1977. *Anal. Chem.*, **49**, 867–70
Rossotti, F. J. C. and Rossotti, H. 1965. *J. Chem. Educ.*, **42**, 375–8
Rossotti, H. 1978. *The Study of Ionic Equilibria*. London, Longman
Sabatani, A., Vacca, A., and Gans, P. 1974. *Talanta*, **21**, 53–77
Sayce, I. G. 1968. *Talanta*, **15**, 1397–411
Sillén, L. G. 1962. *Acta Chem. Scand.*, **16**, 159–72
Sillén, L. G. 1967. *Chem. in Brit.*, **3**, 291–7
Sillén, L. G. and Martell, A. E. 1964, 1971. *Stability Constants of Metal-Ion Complexes*. The Chemical Society, London, Publication Nos. **17** and **25**
Smith, R. M. and Martell, A. E. 1976. *Critical Stability Constants*. New York, Plenum
Sørensen, S. P. L. 1909. *Comp. Rend. Trav. Lab. Carlsberg*, **8**, 1
Sørensen, S. P. L. 1909. *Biochem. Z.*, **21**, 131–200
Sørensen, S. P. L. and Linderstrøm-Lang, K. 1924. *Comp. Rend. Trav. Lab. Carlsberg*, **15**, 40
Sørensen, S. P. L., Linderstrøm-Lang, K. and Lund, E. 1927. *J. Gen. Physiol.*, **8**, 543–99
Stumm, W. and Morgan, J. J. 1981. *Aquatic Chemistry: An Introduction, Emphasizing Chemical Equilibria in Natural Waters*, 2nd edn. New York, Wiley
Vacca, A. and Sabatini, A. 1984. In *Computational Methods for the Determination of Stability Constants*, ed. D. J. Leggett. New York, Plenum (in press)
van Slyke, D. D. 1922. *J. Biol. Chem.*, **52**, 525–70
Westcott, C. G. 1978. *pH Measurements*. New York, Academic Press
Whitfield, M. 1975. *Geochim. Cosmochim. Acta*, **39**, 1545–57
Whitfield, M. and Jagner, D. 1981. *Marine Electrochemistry. A Practical Introduction*. Chichester, Wiley
Williams, D. R. 1974. *Proceedings of the Summer School on Stability Constants*. Bivigliano (Florence), Edizioni Scuola Universitaria, Firenze
Williams, D. R. 1976. *An Introduction to Bio-inorganic Chemistry*, ed. D. R. Williams. Springfield, Illinois, Charles C. Thomas
Williams, D. R. 1984. *Clinical Chemistry and Toxicology of Metals*, ed. S. S. Brown and J. Savory. London, Academic Press

Index

acetic acid titration, 67
Acree, S. F., 31
activity, 1, 3, 5, 23–7
activity coefficient, y, 5
  coefficients, 23–7, 45
Ahrland, S., 82
Allison, S. A., 81
analytical errors, 48–9
anodic stripping voltammetry, 74
Antelman, M. S., 29
Arena, G., 57, Appendix B
Arnek, R., 61
arthritis, 77
asymmetry potential, 13
atmosphere and oceans, 82–4
automation, 2

Ball, J. W., 84
Bates, R. G., ix–x, 1, 9, 16, 23, 28, 29, 30, 31, 32, 33, 37, 38, 40, 41, 42
Beck, M. T., 72
Berthon, G., 71, 75, 77
bioavailability, 12, 74, 76
biological responses, 75
2,2'-bipyridyl titration, 59–60
blood pH, 41, 42
Brauner, P., 44, 45, 61

British anti-Lewisite, 77
British Standards Institution, 1
buffer(s), 3, 33–8
  buffer capacity, 36
  buffer-lines, 69
  buffer solutions, 3
  reference, 9
  secondary, 9

cadmium, 77
Calais, J. G., 42
calcium, sodium edetate, 77
calibration, theory of, 1
calomel electrode, 7, 17
  saturated, 7
cancers, 77
Captopril, 78
combination electrodes, 18
Committee on Safety of Medicines, 75
computer modelling, 75
computer programs, 1
  ACBA, 11, 57–61, 69, Appendix B
  COMICS, 71
  ECCLES, 71
  GEOCHEM, 84
  HALTAFALL, 71
  LETAGROP, 49, 61–2, 69, 71

LIGEZ, 63–8, 70, Appendix D
MAGEC, 49–57, 69, 71, 85, Appendix A
CALIBT, 50–57, 69, Appendix A
MAGEC-MINIQUAD cycling, 52–7, 69
MINIQUAD, 52–7, 71
MINIPOT, 62–3, Appendix C
SCOGS, 71
SUPERQUAD, 52, 71
WATEQ2, 84
concentration scales, 24
Conte, S. D., 64
copper chelation, 77
Covington, A. K., 1, 9, 28, 30, 31, 32, 41
Cremer, 3
Crow, D. R., 21
cyanide toxicity, 74
Czerminski, J. B., 42

daphnia magna, 84
de Boer, C., 64
Debye–Hückel equation, 5, 25–6
degree of acidity, 2
Dempsey, B., 37, 41

deuterium oxide pH (D_2O), 40
dialysis, 74
dielectric constants, 5
Dole, M., 3
D–penicillamine, 77
Dyrssen, D., 11, 76, 83

$E^{\ominus}$, 4
E_{const}, 45–6
edetate, 77
Eisenmann, G., 3
electrodes
  ageing, 16
  bulb resistance, 8
  cleaning, 19–20
  commissioning, 18–19
  design, 13–18
  filling solutions, 16
  glass composition of, 15
  isopotential point, 39
  maintenance, 19–20
  robustness, 16
  specificity, 4
  standard potentials, 23
  storage, 19
  See also combination, hydrogen, ion-selective, reference, silver–silver chloride, special, and stomach electrodes
emf measurements, 2
end points, 46
environmental problems, 84
equilibrium constants, 11
Ethambutol, 78

Federal Drug Authority, 75
Fernbach, 3
filling solutions, 16
Florence, T. M., 74

flotation processes, 81
Ford, M. A., 70
formation constants, 11
Gaizer, F., 62, Appendix C
Gans, P., 49, 52, 71
Gauss–Newton procedures, 49
glass
  composition, 15
  lithium glass, 4
  pyrex, 4
  silica glass, 4
glass membranes, 2
Goldberg, E. D., 82
Goldschmidt. V. M., 82
Glasstone, S., 25
glycine titration, 66
Guggenheim, E. A., 23, 26, 29, 31, 41, 42

Haber, 3
Halstead, B., 76
Hamilton, W. C., 56
Hamilton, I., 42
Hansson, I., 42
Harris, P. J., 81
Hartley, F. R., 61
Hewlett–Packard HP41C calculator, 63 (*see* LIGEZ program)
Högfeldt, E., 72
Huang, Z.-X., 75
Hubert, 3
hypertension, 77
hydrogen electrode, 9, 33
hydrogen ions
  activity, 1, 3, 5
  concentrations [H+], 1, 44–70
  effective [H+] concentration, *potenz* of, 2
  puissance of, 2
  millivolt approaches, 10
hydrogen ions strictly hydronium ions, 4
Hughes, M. A., 3

ICRF 198, 78
ICRF 226, 77
industrial uses of glass electrodes, 79–82
Ingri, N., 71
internal calibration procedures, 47–8
International Union of Pure and Applied Chemistry (IUPAC), 1, 27–8, 31, 38
ion(s)
  ion exchange, 15, 74
  ionic strength, 5, 25
    high ionic strength, 41–2
  ionic strength effects, 45
  single ion activities, 23
ionization constant of water, 6
ion-selective electrodes, 12, 75, 76, 80

James, A. M., 25

Keilland, J., 26, 27, 36
Kelvin, Lord, 3
Khoo, K. H., 42
Klemensiewicz, 3

lead, 77
  lead chelation, 77
  lead nitrate titration, 73
'least squares' approach, 3, 11
Levitt, B. P., 25
Linder, P. W., 15, 27
Linderstrøm–Lang, K., 22, 26, 30, 35
Lindsay, W. L., 84
Linnet, N., 17, 18
liquid junction potentials, 9, 31
lithium glass, 4
Long, F. A., 41
Lund, E., 22

Index

MacInnes, D. A., 3, 26
McDuff, R. E., 83
McGlashan, M. L., 38
Magnusson, V. R., 84
Marcus, Y., 6, 40,
Marsicano, F., 81
Martell, A. E., 72
Mattigod, S. V., 84
Mattock, G., 3
May, P. M., 10, 11, 71, 75
Mead, R., 50
medical uses of glass electrodes, 76–9
mercury halide complexes, 72
medium effects, 6
meters, 20
millivolt approaches, 10–11
mine waters, 84
mixed solvent
  pH, 40
  systems, 25
model calculations, 12
molal scales, 5
molar scales, 5
mole fraction scales, 5
Morel, F. M. M., 83
Morgan, J. J., 83
Murray, K., 27

N,N'–bis(2–aminoethyl)–1,3–propanediamine, 77
Nancollas, G. H., 68
NBS, *see* US National Bureau of Standards
Nelder, J. A., 50
Nelder–Mead procedures, 49
Nernst
  –equation, 23
  –factor, 8
  Nernstian relationships, 45
  Nernstian response, 3
neutrality, 6
Newton–Raphson procedures, 49

nickel, 77
  nickel chelation, 77
non–aqueous solvent
  pH, 40
Nordstrom, D. K., 73
nutrients, 84–5

oceans and atmosphere, 82–4
optimization
  techniques, 11
  theories, 1
oxalic acid titration, 56

Paabo, M., 41
Pantony, D. A., 48
pD scale, 41
perchlorate solutions, 16
Perrin, D. D., 37, 41, 71
pH
  definitions of, 4
  measurement, 22–43
  operational
    definitions of, 3, 9
  pH meter, 7, *et passim*
  pH standards
    French, 30
    German, 30
    Hungarian, 30
    Polish, 30
    Romanian, 30
    secondary, 32
    US NBS, 30–31
  reproducibility of, 1
  standard scale, 3
    definition of, 28–32
    US NBS
      multistandard scale, 9, *et passim*
pit–mapping procedures, 49
pK$_w$, 11
plutonium, 77
potassium hydrogen phthalate, 9
Pritchard, F. E., 25
Prizidilol, 78
protonation constants, 11

Puskas, A., 62,
  Appendix C
pyrex glass, 4
pyridine titration, 58–60

Ramette, R.W., 42
Razoxane, 78
reference electrodes, 1
  calomel electrode, 17
  salt bridge, 17
river-water speciation, 73
Rossotti, F. J. C., 50
Rossotti, H., 11, 50

Sabatini, A., 52, 71
salt bridge, 7
salt effects, 5
Sayce, I. G., 71, 72
Schwarzenbach, G., 19
sea water
  pH, 41, 42
  speciation, 73
silica glass, 4
silicon chip, 1
Sillén, L. G., 44, 45, 49, 61, 70, 72, 82–4
silver–silver chloride electrode, 13, 15
simplex procedures, 49
Smith, R. M., 72
sodium diethyldithio-carbamate, 77
software, 69–70 (and *see* computer programs)
solvent effects, 6
Sørensen, S. P. L., 2, 22, 26, 30, 35
special electrodes, 20–21
speciation, 1, 71–86
  river water, 73
  sea water, 73
  toxicity and, 73
sphalerite, 81
Sposito, G., 84
spot calibrations, 86
stability constants, 11
standard potential, 7, 8, 23

steady state solutions, 75
stomach electrodes, 13
stream water, 84
Stumm, W., 83
synergistic chelation therapy, 77

'taring–out', 8
Taylor series, 62
temperature, 6
tetraethylene-pentamine, 77
titration
  acetic acid, 67
  glycine, 66
  lead nitrate, 73
  oxalic acid, 56
  pyridine, 58–60
  titration equipment, 32, 34
Tomson, M. B., 68
Torrington, R. G., 63, Appendix D
toxicity and speciation, 74
toxic metals, 76, 84
transference effects, 6
triethylenetetramine, 77

tuberculosis, 77, 78

ultrafiltration, 74
US National Bureau of Standards (NBS), 1, 9, 28, 30–31

Vacca, A., 52
van Slyke, D. D., 36
van't Hoff i factor, 23–4

Wang 2200 desk computer, 62 (see MINIPOT program)
water, ionization constant of, 6
Wedborg, M., 82
Westcott, C, G., 7, 17, 39, 41
Whitfield, M., 83
Williams, D. R., 75, 76, 77

zinc extraction, 81
zinc slurry speciation, 82